D1450206

It is the neon sign that blinks on the edge of our consciousness; the wavy, delicate windowpanes in a centuries-old farmhouse; the airy adornment of high-rise architects and playful distraction of daydreaming schoolchildren. Heat resistant or shatterproof, tempered or stained, this magical substance formed of sand and fire has done much more than brighten and beautify: it has changed the very way we live.

William S. Ellis brilliantly whisks readers on a marvelously entertaining journey of ingenuity and discovery, from the birthplace of glass on the ancient shores of Phoenicia to the crystal factories of Waterford, which only recently has leapt into the computer age. In prose as crystalline as his subject, the author celebrates the versatility and functionality of glass, and explains how a substance known to all but understood by few has been shaped and molded to serve mankind in innumerable ways. In these pages, readers will learn how glass has both shaped and been shaped by man's changing relationship to the environment; how it has brought vision to the sight-deprived and to human beings huddling in the dark; and how glass enters the twenty-first century yielding an almost unlimited horizon of possibilities.

With grace, charm and authority, *Glass* delves into history, invention, manufacturing, fine art, and the myriad faces and forms of this protean substance. Whether visiting the flamboyant glass artist Dale Chihuly, dissecting the creation of a twenty-ton telescopic mirror, sampling the history of Tiffany's magnificent lamps, or watching the design and construction of the greenhouses of Kew Gardens, this book treats readers to a multifaceted vision of a material eternally destined to die a violent death, and to be constantly reborn in a relentlessly changing world.

Avon Books are available at special quantity discounts for bulk purchases for sales promotions, premiums, fund raising or educational use. Special books, or book excerpts, can also be created to fit specific needs.

For details write or telephone the office of the Director of Special Markets, Avon Books, Inc., Dept. FP, 1350 Avenue of the Americas, New York, New York 10019, 1-800-238-0658.

GLASS

FROM THE FIRST MIRROR
TO FIBER OPTICS, THE STORY
OF THE SUBSTANCE THAT
CHANGED THE WORLD

WILLIAM S. ELLIS

AN AVON BOOK

Portions of this book first appeared in the *National Geographic Magazine* and are reprinted here with the permission of that publication.

AVON BOOKS, INC.
1350 Avenue of the Americas
New York, New York 10019

Copyright © 1998 by William S. Ellis
Front cover image: Dale Chihuly: Venetian Being Reheated in Glory Hole. Photograph by Russell Johnson.
Back cover image: Dale Chihuly: Translucent Yellow Seaform Persian Set, 1994. Photograph by Rob Whitworth.
Author photograph by Ruth Ellis
Published by arrangement with the author
ISBN: 0-380-79139-0
www.avonbooks.com/bard

All rights reserved, which includes the right to reproduce this book or portions thereof in any form whatsoever except as provided by the U.S. Copyright Law. For information address Avon Books, Inc.

Library of Congress Cataloging in Publication Data:
Ellis, William S.
   Glass : from the first mirror to fiber optics, the story of the substance that changed the world / William S. Ellis
      p.   cm.
   Includes bibliographical references and index.
   1. Glass. I. Title.
TA450.F484   1998   98-21943
666'.1—dc21   CIP

First Bard Trade Paperback Printing: August 1999
First Bard Hardcover Printing: November 1998

BARD TRADEMARK REG. U.S. PAT. OFF. AND IN OTHER COUNTRIES, MARCA REGISTRADA, HECHO EN U.S.A.

Printed in the U.S.A.

If you purchased this book without a cover, you should be aware that this book is stolen property. It was reported as "unsold and destroyed" to the publisher, and neither the author nor the publisher has received any payment for this "stripped book."

 For Ruth

Acknowledgments

In writing this book, I sensed rather quickly that in the end it would read as a celebration of glass, and I remained comfortable with that. My research took me over a good part of the world, each stop a reconfirmation of the universal acceptance of glass as a common material of uncommon reliability and worthiness of praise.

The book evolved from an article on glass that I wrote for the *National Geographic Magazine* in the early 1990's, and it was then that I began to amass a debt of gratitude for what follows on these pages. The *Geographic*, somewhat of a latecomer to belt-tightening, was generous with travel advances and tolerant of outrageous room service charges, and for that I am thankful. Reader response to the article was much more favorable than not, underscoring this truth: people have good feelings about glass for many reasons, such as the material's wide versatility and its role as host for the mesmerizing performance of light in dance. Many people in many places articulated these good feelings for me, and I thank them for that.

There are those, too, who must be named. Ed Linehan, a close friend of many years whose pencil has traced the way to publication for many a wayward text, and Bill Prindle, whose knowledge of glass knows no bounds, read portions of the manuscript and made suggestions for stylistic and factual changes. Stephen S. Power, my editor at Avon, was the first to suggest that this book be done, and through the writing of it his guidelines and encouragement and fine editing kept me going. Jane Dystel, my agent, led me painlessly through contractual thickets, and her otherwise supportive involvement with this project was of importance to me. For assistance with research, I turned to Catherine Fox who found what I needed no matter how elusive it was.

Many of the men and women responsible for lifting gallery glass from craft to art took time to see me and to enthuse about the material as a medium for artistic expression. Sometimes we spoke in the hot shops, and sometimes I was allowed to take a gather of molten glass from the furnace and, under the guidance of the talented workers there, blow a vase or Christmas tree ornament. What I did was certainly not art—not even craft—but those good people in the shops always refrained from giggling, always told me what a fine piece lay there, on the end of the blowpipe. I knew better, but I'd like them to know that I appreciate the kindness.

Contents

x   Contents

GLASS

PART I

BORN OF FIRE AND EARTH

*Who, when he first saw the sand or ashes . . .
melted in a metallic form . . . would
have imagined that, in this shapeless lump,
lay concealed so many conveniences of life?*

—SAMUEL JOHNSON, 1750

Chapter One

In Nature's Forge

Most of the debris on the beach is newly out of the water, having been borne on the morning tide. It is a forlorn place, a palimpsest for the scratchings of shore birds, a small beach where dark, tangled seaweed rims the high-water mark on the sand, and where more than a few dead flounder and whoreson nettles are under siege by flies. The seashells are mostly broken, with nothing among them worth bending for.

There is a vista of a bridge in the distance, a handsome gathering of steel rising high over a wide passage of water, and at night cars can be seen moving across the span in a dance of light. And sometimes there is a show offshore of large fish breaking through the surface in joyful leaps, silvery flashes woven through the spray. But that is not what we have come for, here to the shore of Chesapeake Bay in the tidewater region of Virginia.

The beach where man first made glass: Was it like this, with gulls posted on decaying pilings? Did the Phoenician sailors who pulled ashore there, according to Pliny the Elder, hear the drone of bees around something rotting on the sand, as we do here? Pliny didn't say, but, writing in his

Natural History, he did relate a tale of how the seamen were on the beach preparing to cook over a fire when, finding no stones on which to rest their pots, used lumps of the material they were hauling as ship's cargo. It was a natron, a natural soda used at the time in embalming the dead. When the natron supporting the pots became heated and mingled with the sand, a strange liquid flowed forth in streams, and that, according to the Roman historian, was the origin of man-made glass.

Straight-arrow though he no doubt was, the elder Pliny wandered from the path of truth in the writing of the story. The making of glass by other than natural means predates his first-century A.D. account by 2,250 years or more, and it occurred not along the coast of Phoenicia, what is now largely Lebanon, but in Mesopotamia, or modern-day Iraq and Syria. Pottery makers of that time may have been the first to make glass, but not intentionally. In firing their pieces, they probably fused sand and minerals, resulting in the pottery being glazed with a form of glass. That isn't to say, however, that glass cannot be made on a beach using only the natural materials found there. We are here in Virginia to do just that, to gather some sand and seaweed and shells, and make of them a bit of the material that has been used at least as far back as the time of pharaonic Egypt, when mourners there caught their tears in small glass vials.

As Billy Graham knows scripture, so does L. David Pye know glass. He has come down from Alfred, New York, where he is director of the Center for Glass Research at Alfred University, to collect, prepare, and cook the ingredients. He intends to make glass by invoking the infernal chemistry of molecules beset by fire, all the while following a recipe found in an ancient cuneiform text from Mesopotamia: "Take sixty parts sand, a hundred and eighty parts ashes from sea plants, five parts chalk, heat them all together, and you will get glass." The sand is on hand, of course, though nothing white and fine like sugar, but heavy, coarse. For the other two basic ingredients of glass, soda ash and lime (chalk), he will use seaweed and seashells, respectively. Driftwood will serve as fuel for the fire.

As he cooks the batch, Dr. Pye will no doubt cast himself in the role of one of Pliny's seamen (all of us here feel the nice tug of forces of fancy), and, being a man of good and gentle nature, he is prepared to feign amusement if told he doesn't *look* Phoenician.

At first the fire burns fitfully, but then it takes with a gust of flame and heat. The temperature rises to 1,500°F, and it seems that it will go no higher. The melting process requires a temperature closer to 2,500°, and after two hours, the experiment is curtailed. Still, some of the sand's silica crystals have been transformed into something crude and highly discolored. A person of lesser scientific integrity might say that it was glass, but Pye will have none of that. The specimen is too refractory, he proclaims, and so he will try again, this time with a slight alteration to the recipe.

It is the sand that gives essence to the glass and imparts transparency. While glass can be made from sand and soda ash alone, lime is added to make the glass less vulnerable to attack by water, and to control the viscosity so that the material isn't too fluid. The purpose of soda ash in the batch is to act as a fluxing agent, allowing lower temperatures to bring about the melt. Ordinary bicarbonate of soda in everyday household use can achieve that purpose, as can the natron of which Pliny wrote. On his next try, Dr. Pye substitutes natron's basic compound, sodium carbonate, for the ashes of the burned seaweed, and, lo, after several hours, when the fire has had its way with the batch, he pokes through the embers to uncover a small, roughly textured, blue-tinged piece of glass. It is a lump hardly worth keeping

even as a bauble, but—there can be no question this time—
it is glass.

Dr. Pye need not have gone to all the trouble, since it
takes nothing more than a strike of lightning on a sand beach
to create glass; it appears like crust, in the form of thin (one
millimeter, on average) tubes called fulgurites. Also, when a
fiery meteorite crashes down on the earth, globs of molten
rock are hurled into the air and then fall as small, greenish-
brown bodies of glass called tektites. Among natural glasses,
the most prevalent is obsidian. Shiny and dark, it is born in
the fires of volcanoes; scientists have recovered samples believed to be a billion years old. Obsidian tools have been
found that were chipped and flaked more than a million
years ago. Glass, then, in one form or another, has long
been in noble service to humans. As one of the most widely
used of manufactured materials, and certainly the most versatile, it can be as imposing as a telescope mirror the width
of a tennis court, or as small and simple as a marble zinging
across dirt. It gives us tumblers for drink and bulbs for light,
and, as a mirror, it becomes a scepter for vanity (and lets
us look at the hidden corners of ourselves).

The material is at once heartbreaking and awesomely
strong—a substance with a soul of its own. And, of course,
it is glass that allows the visual union of the outside with the
inside—the glass skins of towers and the windows through
which are reflected the shadows of our secluded lives. It is
a heavy load to bear for something with such a hair-trigger
propensity for breakage, something in which high technology
has only a fairly recent role.

Glass itself is not as simple as the making of it. Although
glass is rigid, and thus like a solid, the molecules of its interior structure are arranged in the random, disordered fashion
of a liquid, a state in which "the ultimate connectivity of the
structure is absent," as Dr. Nicholas Borrelli, a physicist at

Corning, describes it. It was a belief among some scientists for a time that glass consisted of orderly molecular regions that had become entangled as they joined to form the overall structure. That theory was laid to rest in the 1930s when, for the first time, X-ray diffraction of simple glasses showed widespread disorder.

In molten glass, there are silicon atoms and oxygen atoms, all hanging loosely enough to make the glass transparent. The disarray occurs during the melting process, and before the atoms can rearrange themselves in orderly patterns, the glass cools and all movement stops—a freeze-frame of molecular mess. By contrast, the atoms of metals subjected to the same melting crystallize rapidly upon cooling, falling back into formation like troops bellowed to attention.

The computer has value as a tool in research involving the structure of glass, but the fact is, not many important findings or insights come from that (for one thing, there are too many atoms here for a computer to handle). Computers can deal with the theoretical, but the structure of glass needs to be looked at in the format of an experiment, and for that, other techniques, such as those involving the use of X rays and neutrons, are required.

The very looseness of its molecular structure gives glass what engineers call its tremendous "formability," allowing it to be tailored to need. There is glass strong and thick enough to stop a bullet, and glass drawn out in a fiber as fine as a human hair. There is flat glass and hollow glass; ground, cast, milled, and cut glass; and many specialty glasses created to serve the most advanced technologies of today. But even now there are few good tools available to tell us what goes on as the atoms dart around, although glass scientists have started to gain new insight. They suspect that there may be some local order after all in the undisciplined chemical reaction. The order is not far-reaching, but there seems to be enough for researchers to pursue.

Meanwhile, there is glass being made today just as it was two thousand years ago. Sand and soda and lime are still being cooked until the mixture is molten and rude with glaring color. Even the tools are basically the same today as they were then. Makers of handcrafted glass continue to blow through a hollow metal pipe to give shape to the hot gather of glass on the other end. They use crudely carved shaping tools made of hardwood, such as cherry, and a simple tweezerlike spring tool to pull, pinch, widen, and narrow. And like all glassmakers down through the ages, they hold their breath when the blowpipe is tapped to free the finished piece from its end, for it is then, in that brief high season for breakage, that all work can go for naught.

Glass is our most useful, most abundant material, and yet the science of making it remains, after many centuries, to be fully explained. And that is one reason why the span between ancient glassmakers and those of today is without much slack: They share an entanglement in the web of glassmaking mystery. It is why the history of glass evolves as a taffy pull of slow discovery and disclosure.

Lacking scientific knowledge of the chemistry involved in their craft, glassmakers of ancient times must have experimented as they developed the process of core forming, whereby a form made of clay or ceramic was covered with molten glass. When the desired shape was in hand and the glass cooled and solid, the inner core was scraped out. The early craftsmen discovered that by adding small amounts of antimony and manganese to the formula, they could make pieces of glass that were clear, free of the commonplace green hue caused by the presence of iron in the sand. If they wanted pieces with color, such as deep blue, or an opaque cast more yellow than lemon, they found they could do that by adding different metallic oxides to the batch. In that way, they made beads, and vessels—*unguentaria*—to hold the perfumes and ointments used by royalty in the New Kingdom of pharaonic Egypt. Glass held the kohl that accented the compelling eyes of Nefertiti, the creams that gave a glowing trace to the bone structure of her face. And if indeed there came a time when that slender pedestal of a neck was cursed with crenulations of the skin, it was a mirror of glass that brought her the sad news.

Many of the glass pieces made in Egypt at the time of Nefertiti and the New Kingdom have survived to this day, due in large measure to the climate of that North African land. The earth lay dry under the hot, gritty gusts of the khamsin and other desert winds that tumbled across Alexandria and El Fustat (the old city of Cairo), and the aridity served to preserve the glass buried there. Egyptian glass, including inlays of mummy cases and death masks, can be found now in museums around the world.

From the start, the making of glass was tedious, carried out under conditions of ineffable discomfort. In that regard, little has changed in many hot workshops where glass is blown today. The workers are like Eastern mystics as they defy the fire in which the sand and other ingredients have cooked. It is a fire that hardly ever cools down from a level of white, blinding heat, and when a blowpipe is thrust into the platinum crucible to make a gather of molten glass, it is a moment of misgiving, of wondering if it's worth the sacrifice of singed eyebrows. The glassworks is a place of smithylike brawn, a sizzle-and-steam show of hot tools being dipped in water to cool. Among the crafts, none is more shop-driven than the blowing of glass.

Glass was being made for at least a thousand years before the blowpipe was first used, the delay attributable, no doubt, to both the slow progress of metallurgy and the early resistance of glassworkers to change. It has been established, however, that at least one clever Mesopotamian somehow learned to fashion a tube of molten glass, close one end of it and then exhale through the other to form a bubble. The blowpipe itself came on the scene in the first or second century before Christ, and it set off a rich celebration of glassmaking throughout the Roman Empire for the next five hundred years.

It is surprising to learn that China, with paper, gunpowder, and so many other inventions, came late to glassmaking. Glass made there was often fashioned to look like jade or other stones. That was often the role of glass in antiquity: to stand in for precious stones. It could do that so well that it became precious itself. With a value like that of gold, glass at the time of the Roman Empire was a material of luxury and a certain mystery, and maybe even a badge for snobbery. It was certainly that when Seneca, the Roman statesman, was alive and philosophizing, for this is what he wrote: "We think of ourselves poor and mean . . . if our vaulted ceilings are not buried in glass." In certain environments, of course, glass on the ceiling remains in vogue even to this day, but that is another story altogether.

In time there was glass in Rome for all, the poor as well as the wealthy. For the first time in history it fell into common use, as Romans began to drink from glass, store their food in glass, even bury the ashes of their dead in glass. The use of glass cut across class lines, put in service for bacchanal feast or hardscrabble meal. Drinkers favored glass vessels over all others, gold and silver included, and that alone was sufficient to ensure prosperity for glassmakers. In Pompeii there were nearly two dozen wine bars along a road no longer than a city block, and in any one of them, on most any night, men gathered to drink and pass the time in a blur of talk—the hardships of duty in Gaul, boccie, the price of fish, apprehensions about the rumblings from nearby Vesuvius. They took their drinks from glass vessels not yet refined enough to stand upright. No matter: More often than not, they took the full portion in a single lifting and then turned the glass over to rest on its rim.

With glass one could see the wine—sense the taste with the palate of the eyes—and it added to the pleasure of the drinking. Similarly, glass as a panel in the oven allows us to view the rising of a cake, and the cake tastes the better for that.

Drawing on the colossal themes of their times, Roman craftsmen produced masterworks in glass. They paid homage to emperors, and some clandestinely celebrated the advent of Christianity. If indeed there was a seating for the Last Supper, did Jesus Christ drink from a Holy Grail made of glass? There is a vessel of glass to be found in Genoa, in the museum of the Cathedral of San Lorenzo, and believers there will tell you that it is the piece; it is the Holy Grail, they say, the *Sacro Catino*. Genoese sailors brought the dark green bowl back from Palestine. It is broken now, but the reverence for it among many remains intact.

In the 1950s, a man with a passion for ancient glass persuaded officials of the cathedral to let him take a small sample of powder from the bowl. Back in the United States, Ray Winfield Smith took the sample—"a mere crumb of glass," he termed it—to Brookhaven National Laboratory on Long Island, in New York, for analysis. He wrote of the findings nearly thirty-five years ago, in *National Geographic*: "Roman glass of the time of Christ consistently contained close to one percent of magnesium oxide and averaged less than one half of one percent of potassium oxide. But the *Sacro Catino* showed about five times too much magnesium and four times too much potassium to fit the Roman recipe . . . The *Sacro Catino* most probably was made much too late for Christ's hands to have held it." But there can be no challenge to the beauty and aura of the bowl, and for the believers, that apparently is enough to sustain its holiness.

There is one piece of glass from Roman times more celebrated than all the others. It is an amphora called the Portland Vase, and it is often cited as a work of unequaled beauty. It is not known who crafted the nine-and-a-half-inch-high piece, and its date can be narrowed down only to sometime during the two centuries bracketing the birth of Christ. Adding to the stunning appearance of the Portland Vase is the complexity of its construction. It is cameo, dual layers of glass with the outer one—of opaque white, the color of an almond tree in bloom—carved to set off a scene from mythology (that heart-wrenching business with Peleus and Thetis, probably) against a background layer of cobalt blue. Today, cameo glass of as many as four and five carved layers is not uncommon, but such masterful use of the technique two thousand years ago must have been a noble accomplishment.

After the glass was blown for the Portland Vase, it remained for a lapidary to carve the scene, and garlands of eucalyptus should have been placed on the head of that artist. Of course, raging against the extravagances of ancient Rome as he did, Cato the Censor (234–149 B.C.) doubtless would have disapproved had he lived long enough to see the work, since the carved mythological scene is heavy with genitalia. The eared vase was broken in ancient times and then again in 1845, when a crazed museum-goer smashed it with a piece of sculpture. It is whole again now, on display in the British Museum.

In the glory days of the empire there were hot shops for glassmaking scattered over a vast area, from Syria to Spain, North Africa to the north of Britain (strangely, there is no evidence of glass production in Venice during the Roman period, but that would come later, and come with such an impact that the city would reign as the most important center for glass in the world). Massive quantities of glass were produced, especially in the Near East—in a lovely and gentle Lebanon, in sophisticated Alexandria, in Damascus and Aleppo, and on the plains of Babylonia. Some recovered Roman glass suggests that attempts were made to produce panes of flat glass, but with poor results. It was an up cycle for glass in Roman times, a soaring ride to new reaches of use for the material. But much of that ended or changed when the western half of the now divided empire fell for good in the fifth century.

To a large extent, the history of glass is fitted into the blocks of time defined by ruling power and religion. When time ran out on the imperial power of Rome, glassworking began to idle, and another five or six hundred years passed before it was brought to new life, in new colors and shapes and designs, this time by the spreading influence of Islam. From the crucible of the hot shop in Islamic times came glass of new colors, shapes, and designs. Compounds of metals such as gold and copper were applied to the glass and then fired to produce a sheen; luster painting, it is called. Enameling was introduced, and that also involved the firing of a decorative material on the glass—taking a vessel blown of plain glass, and giving it grace and power with stylized representations of birds and fish and other animals, and arabesques painted in gold. The Muslims made glass as thin as a wafer, and glass that was engraved with passages from the Koran in bold, flowing Arabic script, such as this on a lamp that hung in a mosque: "God is the light of heavens and the Earth. His light is as a niche in which is a lamp, the lamp in a glass, the glass as it were a glittering star."

Before there were windowpanes and windshields and pickle jars, the allure of glass, to a large extent, must have been one of pure aesthetics. As soon as the molten glass began to take shape, the work was on a trajectory for the heart; and as the limits of glass were pushed more and more, the artistic merit of the works soared. Those limits were stretched wide during the Islamic period, and the amount of glass produced during those eight centuries was large. In recent years, a team of salvagers working on a single shipwreck off the coast of Turkey retrieved what renowned underwater archeologist George F. Bass described as "a collection of medieval Islamic glass larger than all the world's collections of such glass combined."

It was in the eighth century, when Islamic glass was first being made, that Abu Jafar al-Mansur commanded a hundred thousand workmen to build a city in Mesopotamia. Soon to be the richest city in the world, it was called Baghdad, and it was from there that Islam was ruled for five hundred years. Each excavation for construction in and around modern Baghdad becomes an archeological dig, yielding Islamic glass as well as the gold coins and bronze statues of the past. All of Iraq is like that, a repository for thousands of years of civilization. Indeed, for those with faith enough to wonder not *if* the Garden of Eden existed, but *where,* primary attention should be paid to Iraq, Saddam Hussein notwithstanding.

I went one day to Babylon, *the* Babylon, where I stood on flat, brown ground amid some stones of ruins, and listened as an Iraqi—he looked like Max Baer with a mustache—told me, "This is the place, right here, where the Gardens of Babylon were." I asked him if glass had been found during the excavations, and he said, "Much glass. They have taken glass rods from this ground that were made more than twenty-five hundred years before Christ, and glass that was made after Christ. There is always glass. One day they will find the place here where the body of Alexander the Great lies, and there will be glass with the remains. I know that to be true."

Babylon had been destroyed by the time of the coming of Islam in the seventh century, and by that time, too, glassmakers were again ready to put the material to new tests. Before the end of the Islamic period, blown glass of exquisite design and beauty graced the world. Many pieces survive, but there is one with an appeal more enduring than most. It is a ewer blown around a thousand years ago, probably in Persia. No more than a foot high, it was crafted by covering clear glass scarcely thicker than an eggshell with green glass. The green was then carved to portray a scene of large birds attacking gazelles. The clear glass has now weathered white, but that has lent a dreamy quality to the work. Called the Corning Ewer, with eponymous pride, it is displayed in the Corning Museum of Glass, in Corning, New York. It is among the holdings there that make up the largest collection of glass treasures in the world.

By the time the Crusaders and, later, the Mongols reined up at the gates of Aleppo and Damascus, glass had entered the literature, though not yet as the metaphor it would become. Arabic tracts tell of the mystery—the hints of alchemy—of glass, and sometimes of the lovely thing it can become, such as a vase "which was as though . . . molded of the light of the open plain." Other writings of glass in the Islamic period were meant to reflect on a person's wealth (or pretentiousness), as in "Their baths are large glass houses, heated with cinnamon." For Strabo, the Greek geographer, it was with a gossipy sense of discovery that he wrote, "I heard at Alexandria from the glass-workers that there was in Egypt a kind of vitreous earth without which many coloured and costly designs could not be executed, just as elsewhere different countries require mixtures; and at Rome, also, it is said that many discoveries are made both for producing the colours and for facility in manufacture, as, for example, in the case of glass-ware, where one can buy a glass beaker or drinking-cup for a copper." Clearly, there was glass for both rich and poor.

With the fall of Damascus to Tamerlane and his Mongols, most of the masters there were conscripted for glassmaking service in Samarkand. But the movement for glass production was toward Europe, a route that large numbers of skilled craftsmen from Persia and Egypt would follow.

Chapter Two

Guarded Secrets

I know of a riverbank in the jungles of Sarawak where so many butterflies cluster that they are like a field of wind-blown wildflowers. To come on that scene for the first time is a spiritual encounter, as it is to approach Venice in golden light. And, of Venice, this too: Who has been there and not come away programmed by the heart to remember the city forever? Who has sat at an outdoor table of a café in the Piazza San Marco and not feasted on the glory of that place? Who has been to Venice and not had the shill come close to murmur, sotto Velcro voce, "Want to visit the glass factories in Murano?"

Those who board a launch and travel the three miles across the lagoon to Murano follow a path charted during the fecund centuries of the Renaissance, when Venice-Murano stood as the unrivaled center for glass production. The pieces blown in the shops there astonished the world with their thinness, brilliant clarity, and colors. Royalty throughout Europe stocked their palaces with the glass—not only the vessels, but mirrors and chandeliers as well. The making of Venetian glass was a storied tradition in the history of craft; it flourished and then faded to near oblivion, but it survived, even to this day, when the furnaces there still number more than a hundred.

"Ten glass shops in Murano close each year, and ten new ones open." Fabiano Nicoletti sighed (as the people of Murano are wont to do) when citing the figures. "And there is no industrial area here. The glass factories are mixed in with the housing; you find one factory, a house, then another factory." It is better to ask Nicoletti about the actual making of the glass, since he is director of Stazione Sperimentale del Vetro, a state-run glass research institute. "We are here to help them sort out the main problems of the industry, such as pollution being caused by the gases escaping from the furnaces, and the small, everyday defects in production."

Such hands-on care of the industry by government has evolved from the practice of the thirteenth century, when the ruling doge fussed over the glassworkers with jealous interest. They were as a stable of thoroughbreds, well fed and groomed, but always kept closely reined. The *capitolare,* or book of ordinances, regulating the glassmaking business was detailed and commanded compliance; it was decreed there that cheap glass would not be passed off as precious stone, and, but for the 1750 B.C. Code of Hammurabi, that may have been the first truth-in-labeling law. There were still many secrets to making glass—it wasn't until 1612 that the first textbook on glassblowing, *L'Arte Vetraria,* by Antonio Neri, a Florentine glassmaker, was published—so industrial espionage was a cause for concern. The secrecy, that inbred sense of withdrawal and steely aversion to change, remained a trait of Venetian glassmaking for centuries.

It is a simple matter now to determine the components of a piece of glass, and so few mysteries remain about the compositions and colors of Venetian glass. But as recently as thirty-five years ago each factory had its own closely guarded formula, not only for making glass, but for pollution control and conservation of the gas that fueled the furnaces. To this day there remains one shop where the formula is as secret as it was hundreds of years ago. The room where the raw materials for glass are prepared is partitioned into two sections. Workers in one section place a quantity of each material on a scale, but cannot tell its weight. The scale's readout lies beyond the partition, where the shop's owner yells a full-throated *"Basta!"* when the scale registers, for his eyes only, the precise numbers called for in the recipe. It is not surprising to find the work of some Venetian glassmakers—those who experiment with new formulas—jealously guarded, for have they not long been known as *l'uomo di notte,* men of the night?

Of course, it is all an open secret, but the pretense continues, paced in its long, long run by both the extravaganzas of glassmaking and Venice itself as brilliant eulogist for the past.

Near the end of the thirteenth century the glass factories were relocated to Murano; so many furnaces posed a fire hazard to Venice. On the smaller island, too, the craftsmen were less likely to reveal their secrets or be lured away to foreign shops. In time they would be forbidden to leave Murano on penalty of death (assassins were sent to track down and dispatch any artisan who attempted to market his skills outside of Murano), although one determined to slip away could manage it. Yet in some regards they shared levels of respect with nobility. Under the banner of their guild, they paraded proudly on feast days and other celebrations.

For the most part the glass blown by the craftsmen of Murano was light and clear, gossamer to the touch, perfect for use at a papal mass or to hold the unguents of a contessa. The secret lay in a new glass they called *cristallo.* The breakthrough was achieved with the use of a vegetable ash rich in potassium oxide and magnesium. The *sale di vetro,* or glass salts, extracted from the ash were used as the fluxing agent in the melting process, mixed with the sand to produce a sodium-potassium-based crystal-like glass without the weight of today's lead crystal, but with equal clarity and elegance. *Cristallo* responded superbly to the breathy commands of the blower, moving to form without fuss.

By the start of the seventeenth century, Venetian glass covered a wide range of forms and designs. There was glass with delicate enameling, jeweled glass with gilts and gleams of silver, clear glass latticed with opaque threads, glass of soft colors (and sometimes not so soft), the so-called ice glass with its shimmering, shattered appearance, obtained by dipping the still-hot glass in cold water. The goblets they made were daringly thin and fragile, unlike the imposing chandeliers. Once again, glass as the most pliable of materials had been stretched to new limits, as it would be time and time again, and as it continues to be today.

Venetian glass through the Renaissance and beyond was the finest glass made in all the world, and most of the world—Europe, especially—was envious. Some of the skilled glassblowers of Murano made their way to Spain and north to Germany and France, England and Bohemia, and the industry took root in those places. The Venice-Murano grip on glass production began to loosen, and by the middle of the nineteenth century, production there had all but ceased. Tastes had changed, with a new preference for the heavier lead crystal perfected earlier by an Englishman named George Ravenscroft. He had added lead to the materials mixed for fusing, and that produced a glass without the clouding often caused by the fumes of alkali in the mixture. It was glass of brilliant clarity, and well suited for engraving.

Remember the time not long ago when it was something of an article of faith that a family should have at least one nice piece of crystal in the china closet? It wasn't cheap, but not painfully expensive, either, and it had a good feel to it, heavy—at least a 24 percent lead content—but sensual. The patterns were cut into the glass, and if you tapped it with a fingernail, the sweet ring resonated through the air. And how grand it was when light struck the piece, when the lead caused the light rays to further refract, or bend, separating out the red and green and blue to give sparkle to the glass. Such was the cut crystal, mainly from England and Ireland, that banked the fires in the furnaces of Murano.

Although it wasn't long before Venetian glass was once again in production, the push for new procedures, better furnaces and fuels to fire them, and new families of glass was most intense elsewhere in Europe. In Bohemia, there were close to a hundred shops where glass was made, much of it the brilliantly colored glass with which that land, later to be called Czechoslovakia (and still later to be divided into the independent Czech and Slovakia Republics), has lastingly identified. Of course, the industry in Bohemia was active as far back as the 1300s, when stained glass with elaborate designs was among the many works being crafted. That artistic tradition has continued through the years. Indeed, creative glassmaking in Czechoslovakia is a cultural icon of equal importance with painting and sculpture and architecture. Even now, when glass art is taking leave of the world of craft and ascending into the price-upon-request cachet of galleries and glossy catalogs, the heirs of the Bohemian tradition remain in the top strata of talented glassmakers.

In England, a towering cone-shaped structure arose in the West Midlands town of Wordsley, near Stourbridge, to hold a great furnace with twelve melting chambers. The time was the early eighteenth century, and it had long been decreed that wood could not be used as fuel for the fires to melt glass. It took ships to rule the seas (as well as to transport smuggled opium) and it took timber to make those ships; it is a historical given that, in England, the launching of a man-of-war with fifty guns took precedence over the blowing of a stemmed goblet. With no other choices, then, glassmakers turned to coal, and quickly found that a coal fire required a draft of extraordinary force to produce enough heat to melt glass. And that was the glory of the ninety-foot-high Redhouse Glassworks cone in Wordsley: The air could be sucked in through the bottom, while hot gases shot out through the top, creating a fine draft; the cone was, in effect, its own flue and chimney. Glass was made there for more than 125 years. More than that, the glass industry in that area of England would gain wide recognition, not only in Europe, but elsewhere in the world as well.

Fifty tons or more of coal were fed each week to the dozen crucibles of the furnace. The work force numbered about fifty. It was hot and dark and noisy in the cone, but that could be as much blessing as curse during the cold and damp winters, a time when even the gentlest of those good people of Wordsley turned fractious. The men worked six hours on, six off, around the clock, but for the women there the shift was longer. They sat at long benches for up to nine hours, scratching designs through the wax that covered some of the glass. The glass was then dipped in acid, etching the design onto the surface of the piece where the wax had been removed. It was plain sheet glass, however, that led production in the early years of the operation, followed later by bottles, including medicine vials, and crystal tableware. Stuart & Sons Ltd. continues to manufacture high-quality crystal at the site, although the two-hundred-year-old structure is in use now only as a museum.

There is a room in the cone where grown men no doubt once wept. Christine Colledge, curator of the museum, explained that the tax collector encamped there, weighing each piece of glass as it came out of the annealing oven, or lehr, where it cooled down. The tax, called the "shrower," was levied on weight, and it was a heavy yoke on the English glass industry for a hundred years. Just making the sheet glass was challenge enough, for it was a time when such production was without the dissolves and zooms of innovation and progress. To make such glass, a craftsman at the Redhouse cone had to blow a cylinder of molten glass as close to flat as he could get it. The cylinder would then be split lengthwise, and both pieces heated and ironed flat. There would usually be bubbles, or "seeds," in the glass, along with other imperfections such as tints of color.

Today, there is at least one glassworks where all glass is mouth-blown, and where flat glass for windows is made just as it was at Redhouse. Other than for restoration projects, such glass is in small demand, but there is one large structure in need of it: the White House. Government code requires that the panes of the windows in the 132 rooms of the mansion be of mouth-blown glass, blemishes and all. Panes for the windows were last purchased in the 1940s, and the stocks have long been depleted. "They didn't even have a spare piece we could use as a model," Richard Blenko said. "They've been getting by on patchwork."

Blenko Glass, in the small West Virginia town of Milton, will be making new/old glass for the White House, and there are few, if any, shops in the country as qualified to do the work. The company, now in its fourth generation of Blenko family ownership, made the fourteen thousand pieces of stained glass now in the chapel of the U.S. Air Force Academy at Colorado Springs, as it did all of the original restoration glass for Colonial Williamsburg, as well as the art deco glass for the restoration of New York City's famed Rainbow Room nightclub.

There is a good feel to Milton, a peaceable place and a refuge from the turpentine culture of West Virginia's hills. Even the Blenko shop has been resuscitated—somewhat, at least—from the fiery, noisy netherworld atmosphere found in almost every glassmaking operation. There are "gaffers," or master glassblowers, there who have been with Blenko for as long as fifty years, one of whom, when we spoke, was crafting a glass obelisk that would be presented on television as a Country Music Award. At the same time, another, younger gaffer on the floor had just lost a bowl-in-the-making to breakage, and as I watched, he slumped a bit, like a miler at the finish. He was successful on the second try, however, and it was a handsome piece he blew, asymmetrical in design and with all the cherished imperfections of hand-blown restoration glass. Running to just under a foot and a half in length, it would serve perfectly, it seemed to me, to hold some pears and grapes.

The government is expected to order twelve hundred of the eighteen-by-thirty-six-inch panes, enough to last for twenty years. They wanted the glass to have a yellow tint, but Blenko's people talked them out of that. Tinted or not, it will be glass with a tradition, and in the United States, that tradition goes all the way back to the founding of Jamestown.

The oldest English-made glass in the United States was not found at Jamestown, though. In 1991, archeologists working at Roanoke Island's Fort Raleigh National Historical Site in North Carolina uncovered some shards of English glass more than four hundred years old. The pieces were found in the ground near the site where the legendary Lost Colony encamped in 1587. There were other remnants found with the glass, such as equipment to distill water, pieces of burned brick, and other artifacts, indicating that a laboratory for metallurgical studies existed at the site, meaning, of course, that it was here that scientific research in the country began, long before the inventiveness of Thomas Jefferson and Ben Franklin. Historians, as far as I can tell, were not electrified by the find; they were hoping no doubt for clues to the fate of the men and women of the Lost Colony. But there was nothing, not a cryptic map scratched in whalebone, not a button or rusted sword hilt, just shattered remnants of their existence.

Among the first colonizers who arrived in Jamestown in 1607 were half a dozen or so glass blowers, most of them Germans. They came to stay in this, the first permanent English settlement in continental North America, and they came to make money, to return a profit to investors in the Virginia Company of America, chartered by King James I. The seeds they planted were not only for crops, but ultimately for a new nation. But Jamestown was a bear for those who held shares in the sponsoring company. There was disease and hunger there, and death. The glassblowers among them moved into the piney woods, a mile from the settlement, where they managed to build a furnace and start making glass. They made bottles, and they also made beads. Glass beads often could defuse the hostility of the natives, as Columbus found in the Indies when, as he noted on one page of his log, he gave some of the trinkets to the Indians and in return "got a wonderful hold on their affections." That endeavor in Jamestown also failed, but not before it established glassmaking as the first industry in North America.

It wasn't until 1739 that the first *successful* glass factory was established in the New World. Defying English policy prohibiting all manufacturing in the colonies, Caspar Wistar chose Alloway, N.J., as the site for his glassworks, and there it would operate for thirty-eight years.

By that time, more German glassmakers had crossed the

ocean to begin the slow process of building the industry for a productive and lasting presence in the Americas. They fired their furnaces in New Jersey and Pennsylvania, where there was coal for fuel and sand and other raw material for making the glass. By then, there was a heavy need in the colonies for window glass, and so that was crafted in large quantities, along with tableware and other household needs of a nation in the making. And the glass was getting better—more refined, more attractive. But that wasn't good enough for Henry Stiegel, a German glassmaker from Cologne. Thirteen years after his arrival in Philadelphia in 1750, Stiegel opened glassworks at Elizabeth and Manheim, both in Pennsylvania, and before long he expanded and brought in master craftsmen from Venice and England and other parts of Europe. In addition to bottles and blown table glassware, he made lead glass of a quality to compare favorably with that made in England.

The English, burdened at the time with taxes on almost everything, including windowpanes (there remain in England today some very old houses with framed windows bricked up rather than glassed in), were not of a mind to wish the new industry well. If anything, they viewed it as obdurate behavior by the colonies. For glass made in the colonies reflected growing discontent with Mother England. Each tumbler the settlers made, each bowl and wavy windowpane, was a strike for self-sufficiency, an emollient for the chafing over import duties imposed under the Revenue (or Sugar) Act of 1764. They would have to pay the levies on molasses and calicoes brought to those sunrise shores, but the glass, for the most part, was from within, born of the New World's sand and fire.

With the end of the Revolution, glassworks in America pushed their roots even deeper into the economy and enterprise of the newly independent nation. In the 1820s Bakewell,

Page & Bakewell, in Pittsburgh, and the Boston and Sandwich
Glass Company, in Sandwich, Massachusetts, started to make
pieces of glass using mechanical pressure. Previously, small
and simple pieces such as seals and weights were made by
using a handheld implement to press molten glass against the
sides of a mold, but at Sandwich and Pittsburgh they did the
pressing with a fully mechanical plunger. From the original
patent, others followed, detailing improvements in the pressing
forces; unfortunately, those records were lost in an 1836 fire
at the U.S. Patent Office in Washington, D.C.

Ordinary glass for the first time was free of the time-
devouring restraints of handcrafting. Production soared,
prices fell, and items of glassware came into household use
throughout most of the world. While the need for skilled
glassblowers declined, the work involved in making this new
glass was nonetheless exacting. The fire had to be maintained at a certain temperature so that the molten glass was
not red hot, but almost so. The amount of the gather put
in the mold had to be equally fine-tuned, for if there was
too much or too little, the glass would be flawed. The hinged
molds were made of metal—brass or iron, usually—and they
were of such wide-ranging sizes and shapes that the product
lists of the shops seemed endless. There were molds to make
pieces for use not only in the kitchen, but all over the house.
They were pressing glass for doorknobs and inkwells and
fittings for furniture, for example, and nightstand pieces to
hold the jetsam and flotsam of pocket and purse.

Pressing glass mechanically was the most important development in the history of the material since the development
of the blowpipe. Such a departure from the ordinary was
rare in glassmaking, and it would remain that way until the
early years of the twentieth century. Since then, most major
achievements in glass technology and production have occurred. However, there is much prewar glassmaking machin-

ery still in operation—apparatus of an age and quirkiness to be almost endearing. There is also a strong and enduring loyalty to tradition, and nowhere is that more evident than in the otherwise unremarkable town of Waterford, in the south of Ireland. It is there that the most highly prized mouth-blown crystal in the world is made. In straddling the past and present, no one in the glass industry does it with greater ease than the crystal maker.

"It's all romantic, but rather costly," Redmond O'Donoghue, chief operating officer of Waterford Crystal Ltd., told me at a time when the company faced financial hardships (it has since turned around), speaking of the shops and the aura that hangs in them like a thick fog. Waterford Crystal is a place that inspires thoughts of ancient guilds and modern confrontational unionism, of the passing of craftsmanship from father to son, of an unqualified pride in work, and a somewhat sweet indifference to profit making. There is the thought there, too, that the punishing heat of the furnaces is not to be damned, not as long as there is an Irishman at Waterford with memories of a winter in childhood when there was no coal in the bin.

It is the same in southern Sweden, in a vast region of forests and lakes and winters that chill to the bone. There Orrefors crystal is made—pieces of power and lyricism—and where the past is not forgotten. In bygone times, workers left at the end of the day and then returned with their families to have supper in the warmth of the furnaces. And that tradition, called the *hyttsil*, lives on today; in a Swedish winter, the glass hot shop remains a cozy place to have your herring and potatoes.

Management treads lightly on that path between then and now. Øyvind Saetre, director of marketing for Orrefors, said this: "It is a task facing management to run the company as a business without destroying the mystique." In Waterford, it was not enough that the company makes twenty-five hundred different glass products, ranging from chandeliers to paperweights, that it was a Waterford chandelier under which the Declaration of Independence was signed in Independence Hall, in Philadelphia, that it is Waterford crystal that heads of state give to other heads of state. It didn't matter that in a poll taken early in the 1990s to determine luxury goods favored by Americans, Waterford crystal placed second, topped only by Samsonite luggage.

What mattered was that in the early 1990s the demand for high-quality, mouth-blown crystal had dropped. In part it was a matter of fashion and popular taste, and of the fact that market was being hammered by world recession and the decline of the U.S. dollar, a crucial factor in that about 60 percent of Waterford's volume is exported to America. "When the dollar is strong, business is terrific," O'Donoghue said at the time. "But when it weakens, we have problems." In addition, Waterford then was working to combat fears of possible health dangers from lead in glasses and decanters.

The answer was to do what traditional glassmakers are reluctant to do: make some rocking changes. Waterford announced that it would start making a cheaper line of crystal, not in Waterford, but in Germany and parts of what was formerly Yugoslavia. It would be lighter crystal, of more modern design, and with added appeal to younger buyers. They would call it Marquis. The purists decried the move. Nevertheless, the company went ahead with the new production, and the success of Marquis has pumped new life into the firm.

Waterford went a step further. In installing equipment in the new shops abroad, the company chose modern computer-controlled electric furnaces, an ocean-span leap from the pot furnaces of eighteenth-century design in the home plant. With the new furnaces came fewer imperfect pieces; the rate of rejection dropped, along with costs.

In the face of all that, Waterford continues to make the crystal it is famous for—the classical, heavy, deep-cut glass of a pedigree all its own. The company is too much of a giant in the world of top-quality crystal to give up that market. The old machines run on and on, and each work produced in the shops at Waterford continues to be handmade, each fractionally different. And there are no "seconds" at Waterford. An imperfection, no matter how small, is cause for breakage, and so there are workers there who spend a good part of their day smashing Waterford crystal. Being invited to participate, I was handed a bowl with a bubble in the glass; but for that, the piece would have been priced in a store at $200. And so it was mine to destroy, to hurl and shatter, and I could laugh depravedly while doing it if I chose. The bowl carried into the side of a metal bin with force enough to set up a good noise, part explosion and part crunch, followed by a rain of joyful and most satisfying tinkling.

Glassmaking as practiced by Waterford and other makers of fine crystal, such as Orrefors and Corning-owned Steuben, is decidedly low-tech. It is men and women with high skills, blowing glass, lovingly cutting it, polishing and engraving, squeezing the last turn of perfection from the handwork so that light becomes the music of the crystal piece— the dance of it through a chandelier, for example, or the soft sparkle of colors in and about a bowl.

And that's the way they did it many centuries ago. That's the way glassmakers in Murano made the world catch its breath over their work. Of all that beckons along the trail of the early history of glass, nothing does so as compellingly as Venice and Murano. And so I returned, this time in the wake of heavy rains. The Piazza San Marco was underwater, but a group of musicians on a stand outside one of the cafés continued to play; a Venetian friend called it "music to sink by." The passengers on the launch to Murano were stern-faced and silent, as if waiting for the fire of heartburn to subside.

Today in Murano, much of the glass produced is for the tourist trade. As much as 40 percent is exported. For the most part it is not the Venetian glass of renown, but there are some carryovers from the golden age still in place, still making glass of high quality and innovative design. Among them is Venini, a glassworks largely responsible for the first resurgence of Venetian glass in this century; it began in the early 1920s and lasted for about ten years, followed by an-other decade of prosperity starting in the 1950s.

The founding proprietor was Paolo Venini, a lawyer and an entrepreneur with a love of glass. He named the company Vetri Soffiati Muranesi Cappellin-Venini e C., and the method of operation there was a departure from the past. He employed designers to come in and work with the glass-blowers, resulting in the making of glass touched by the genius of Italian form and color. It was glass of luster and elegance, and of a dazzling palette; it was glass to put the world on notice that Venice-Murano was back on top. Ven-ini let fresh air into his shop, defying the tradition of con-finement for the sake of secrecy. Designers from many lands, eager to work with glass, made their way to the island. And when this latest of the boom times for Venetian glass ended in the 1960s, there were Venini pieces marked for everlasting recognition, including the vase with the design of a standing folded handkerchief.

Vetri Soffiati Muranesi Cappellin-Venini e C. (the e C. stands for "and Company") remains in business. "I'm not sure if we are the largest factory now on Murano, but we are certainly the most famous," said Lucio Moretti, a Venini spokesman. "We make about a thousand different glass products, including chandeliers. The price range? It is all expensive."

As there is a Venini with its 150 workers and twentieth-century approach to glassmaking, so too is there Alfredo Barbini, a maestro of both design and glassblowing who works in the old tradition. He is an elderly man afforded uncommon respect by other glassmakers, having spent more than three score years creating sculpture in glass. His signature pieces have helped to preserve the tradition, as did the works of the Barbini family before him, going as far back as 1638. He is now eighty-three years old, but he knew even as a boy what he wanted to do with his life. "I decided I would rather work with glass than go to school," he said. "So I started when I was thirteen. Two years later, I was regarded as a master."

Barbini is still at the furnaces on most days. He is short and wiry, and as he handles the blowpipe with the heavy gather of molten glass on the end, the skin of his arms and shoulders is pulled tight to reveal more muscle than a man of eighty-three can rightfully expect to have. His demeanor is imperious in the style of master glassblowers, for it is he who understands the moods of cooling molten glass better than anyone else. The work must be done quickly, the movements without waste, the creativity fitted into time slots defined by flash chemistry. And through it all, the great uncertainty is always there: Will the piece survive the arrant threat of breakage? He is surrounded by the thirty workers in his studio, and I am close, too, to watch him work and see a fish take form in glass—glass that he reheats and reworks time and time again until it is, in his eyes, as perfect as he can make it. When it is finished and taken to the annealing oven to be cooled down over a period of three to four days, there is a respite of no more than an hour before Barbini has started on something new with the same nearly spiritual approach to glassmaking that has been the mark of masters since ancient times.

I met an aged glassmaker in Murano, a *padrone* of the furnace, who spoke of his burn scars as if they were chevrons awarded for lengthy service, for even the masters cannot always escape the searing heat that blows out when the opening in the furnace—the "glory hole"—is uncovered. He, like his father and his father's father, will complete his working life in the hot shops, crafting some pieces approaching heavenly beauty, although, as he said, it is the face of the devil he sometimes sees in the mean fires of the furnaces.

But glass has muscle as well as soul, and at the same time Venini was getting his factory in operation, architects, engineers, scientists, and businessmen were looking at glass with an eye for function rather than design alone. They were reaching into that well of versatility, and it was deeper than they had ever imagined.

Chapter Three

Bulbs and Bottles
(and Breakage, Too)

It can happen at any time: an inspiration, a revelation, an idea that flashes to life and then burrows into the mind. In 1922 it happened to a man named Will Woods while he was working as a foreman at the Corning plant in Wellsboro, Pennsylvania. He envisioned a machine that would make light bulbs automatically, with no hands-on involvement. Prior to that time, bulbs were mouth-blown, and later made by a semiautomatic process that required a good deal of manual assistance. Peak production at any one operation was forty-two bulbs a minute.

The cost of a light bulb at the turn of the twentieth century was equivalent to half a day's pay for the average worker. (Even so, it was a golden bargain in at least one case. Today, in a firehouse in Livermore, California, a light bulb continues to burn as it has since 1901. It is a bulb of mouth-blown glass with some ceramic in its base. Giving three watts of power, it shows a dim red, good for use, according to Tim Simpkins of the fire department, as a night-light.) In effect, Will Woods's invention lighted the nation. The price of a bulb plunged, and the pale yellow light of those early filaments flickered in households from New York to San Francisco. Although the machine has been called one of the great mechanical achievements of all time (it was designated by the American Society of Mechanical Engineers as the tenth international landmark in mechanical engineering), its success has been measured by its enormous contribution to the common good.

Today, 1.8 billion light bulbs are manufactured each year in the United States, and it is that same type of machine that makes them. Called the ribbon machine, it has not been surpassed, and yet it is invested with nothing so much as wondrous simplicity. A narrow ribbon of molten glass travels over a moving belt of steel in which there are holes. The glass sags through the holes and into waiting molds. Puffs of compressed air then shape the glass. In that way, the envelope of a light bulb is made by a single machine at the rate of sixty-six thousand an hour, faster than a speeding train clacking over the rails.

The ribbon machine has drawbacks. "For every ten thousand pounds of molten glass that goes into the machine, only thirty-five hundred are used," said Jeffrey Hoffman, a former operations manager at the Wellsboro plant, now owned by Osram Sylvania. The wasted glass is called cullet. Some of it is reused in the furnace, and some of it is combined with other materials, such as ceramics, to make products like glass-based tiles.

Corning no longer makes bulbs for general use, but its original ribbon machine lives on at the Wellsboro facility. There have been many improvements to the old machine, but it still functions with all the clanking, clunking, and hissing of before, and with all the parts that glow with heat and smell of grease. To watch it in operation is like witnessing the adumbrated dawning of doomsday. But the bulbs tumble onto the conveyer belt by the tens of thousands, and seeing that, you might want to applaud the breakage rate of only between 3 and 4 percent. Or you might wonder, as I did, that seven thousand bulbs, made of glass no thicker than twenty thousandths of an inch, can be packed naked in a wooden crate with no protection against breakage other than a light coating of lubricant on each.

The plant at Wellsboro, in the scenic hills where Pennsylvania and New York come together, has had a long and deep involvement in the use of glass coupled with light. The original factory dates back to 1870, when window glass was made there. At the turn of the century, when the country was moving from gaslight to electricity, the production of mouth-blown bulbs began. Later, Wellsboro would make over two hundred thousand radio tubes a year, before solid-state technology outdated every parlor's Philco or Atwater Kent. Today, in addition to light bulbs, the plant is a major producer of lamps for automobile turn signals, for overhead lights in airplanes, and even for the little light on the vacuum cleaner.

Since that day more than a hundred years ago when a team of glassblowers answered Thomas Edison's request and produced a bulb that would work in the great man's electric lamp, the union of glass and electricity has gone on to be of essential service in some of the most advanced medical procedures, as well as in the new age of communications and the movement of information. It is also a strong force for innovations in the design and function of buildings, especially urban towers.

One coupling of glass with electricity raised a racket in homes across the country, and that was in 1924 when Corning began making glass insulators for power lines. As a replacement for porcelain, glass was ideal except for one thing: It caused static in radios when a buildup of electricity on the glass crackled and popped. Coming over those early airwaves, it made the Happiness Boys broadcast, out of WJZ in Newark, New Jersey, sound as if it was originating in Addis Ababa. The problem was solved by coating the glass with tin oxide, allowing surplus electricity to drain silently away.

Understand, electrical current is not in unattended flight through the glass itself, as the resistance of the material prevents it from being an effective conductor. Inside Edison's light bulb was another piece of glass in which the current-carrying metal electrodes were sealed. The virtuoso performance of glass when acting with electricity is in the embrace of the light itself, be it the light of a laser in a glass fiber or that in the ordinary light bulb—the standard, screw-in light bulb that has forever figured into the derisory equation concerning X number of Polish people.

Even before Edison's discovery, glass was being used in chimneys for lamps fueled by whale oil and, later, kerosene, because of the added brightness the material provided. Indeed, when Ami Argand, a Swiss, devised a lamp with a tubular wick to feed air to the flame, and with both a chimney and a shield made of glass, the intensity of the light was such that American-born Count Rumford (Benjamin Thompson), British resident physicist and statesman, cautioned in 1811: "No decayed beauty ought ever to expose her face to direct rays of an Argand lamp."

The demand for bottles was no less pressing than that for bulbs. For one thing, the ladies of the temperance movement were going around smashing bottles, in addition to scaring people half to death with their chilling chants: "The drink, it paints men's noses red, and blights their lives, and kills them dead." They pushed bitters as a substitute for liquor, and bottles were needed for the dozens of brands that went on the market. And of course there was a need for bottles for medicines and other things, as well as for glass containers of various sizes and shapes. Bottles were still being made with human lung power, by blowing molten glass manually into a mold, and then shaping the stem and mouth of the piece by hand. It remained for someone to invent an automatic system, and that is what Michael Owens did in 1903 while employed at the Libby Glass plant in Toledo. It was a machine that replaced the arms of the glassblower with mechanical arms that rotated around a vertical axis, and it substituted compressed air for his breath. From that time on, commercial glassmaking, for the most part, was out of the skilled hands of the master blowers.

But there was still time for a young writer to observe the old way of making bottles, and this, in part, is what Carl Sandburg had to say in 1904, in "In Reckless Ecstasy," the first of his writings to be printed:

Down in southern New Jersey, they make glass. By day and by night, the fires burn on in Millville and bid the sand let in the light.

Millville by night would have delighted Whistler, who loved gloom and mist and wild shadows. Great rafts of wood and big, brick hulks, dotted with a myriad of lights, glowing and twinkling every shade of red. Big, black fumes, shooting out smoke and sparks, bottles, bottles, bottles, of every tint and hue, from a brilliant crimson to the dull green that marks the death of sand and birth of glass.

From each fire, the white heat radiates on the "blowers," the "gaffers," and the "carryin'-in boys." The latter are from nine to eighteen years of age, averaging about fourteen, and they outnumber the adult workers. A man with nothing hailing from nowhere, can get an easy job at fair pay, if he has boys who are able to carry bottles—many men in Millville need no suggestion from Roosevelt—boys can carry bottles and girls can work in the cotton-mills near by.

Of all the glass containers ever made, none seems to stir such sweet memories as the old milk bottle. For many, it has become a metaphor for lost youth. Those on the downward side of life remember the milk bottle the way they remember what they were doing on that April day in 1945 when President Roosevelt died in Warm Springs. Visibility surely played a role; the milk could be seen clearly in the bottle, as white as Bismarck in winter, cream resting on top beneath the waxy paper cap, frozen like a piece of vanilla Popsicle if the milk was left out too long on a winter's morning. And when the milk was finished, the bottle lived on, sometimes to go to a party where it was spun to determine who would get kissed.

Today milk is still sold in bottles, but usually only as a specialty item in a pricey store. Sadly, the true epitaph of the container's classic design may be nothing more than the battered wooden bottles of the baseball toss at the carnival. Plastic is now the container of choice for milk, as it is for many items once showcased exclusively in glass. There are advantages to using either material, but the one that carries the most weight has the least. Being lighter, plastic is winning the competition.

But glass holds some aces, too. Because the material is inert, odorless, impermeable, and nonporous, there is no abuse to the flavor of the product it packages. Glass retards spoilage (drinkable wine has been recovered from shipwrecks more than a century old) and therefore offers shelf life far beyond that of plastic.

And then there is the image factor. Glass is far and away the first choice for packaging of upscale food and drink products. Can you imagine Absolut vodka in plastic? Not likely, not now anyway.

In Europe, new and trendy products are almost always premiered in glass. That even extends to moist pet food, such as Festin de Morceaux, an all-meat feast for the cat (as a pet food that dares display itself before purchase, it must have merit). And although peanut butter is rapidly going plastic, the first of the spreads to be horizontally striped with strands of hazelnut chocolate made its debut in England in glass. But in the broader picture, polyethylene, having joined with aluminum to capture the soft-drink container business, is in command. Although making and selling glass containers remains a $4 billion business, that is far less than its counterpart in plastic. The number of facilities where glass containers are made has dropped 20 percent over the past twenty years.

Although the major consumer of glass containers is the beer industry, aluminum is moving toward what is likely to be irrevocable command of that packaging, with a market share now of 45 percent as compared to 35 percent for glass. Again, weight and compactness are factors. It's not easy to dangle a six-pack on a finger when the beer is in bottles—those familiar amber-colored bottles made dark to absorb ultraviolet light and therefore protect the brew from developing a "skunky" taste. For my money, the cold kiss of sweaty metal is no match for the gentle pucker of a bottle's lip, and of course there can be no question that genteelness favors glass over the aluminum beer can, with its shadowy threat of some burping and brawling. Knowing that, the makers of beer commercials, such as those shown during the televising of Super Bowl XXXI in January 1997, almost always use bottles to showcase the product.

Alas, the glass beer bottle, like that in which milk once came to our tables, cannot but vanish before too many years pass. A pity. That does not mean that aluminum will have a clear field. There is a new plastic known as PEN, capable of withstanding heat of 140°F, necessary for the pasteurization that gives beer longer shelf life. But as of now, PEN is too expensive for such commercial use.

It is in the moldable nature of glass to give the user what is needed without undue complications. To make the glass amber-colored for the beer bottles, as an example, all it takes is the addition of carbon and sulfur to the formula. For deep blue, add cobalt; calcium phosphate for opaque white; gold for red or ruby; antimony for canary yellow. For a time there were glasses of a strange yellow-green hue being made. But the color vanished from the market during World War II, when the atomic bomb was being developed, no doubt because the yellow-green was obtained by adding uranium to the glass.

Colored or not, much old glass goes back for another burn in the furnace. For every hundred pounds of silica sand that go into the mixture for melting, there is an equal amount of cullet, or used glass, that has been broken up into small pieces. There is a lot of that available—more than 30 percent of the nearly ten million tons of glass containers produced yearly in the United States is recycled. On average, close to 30 percent of a glass container is cullet.

The glass container recycling program began in 1975 as an alternative to the returnable-bottle system. There are bottles yet that are simply cleaned and reused, with an average life span of seven refills, but for the most part the time has passed when a kid in need of movie money could scavenge the neighborhood for refundable bottles. But it's still done in many third-world countries—in the famine-stricken Sahel region of sub-Saharan Africa, for example, where sometimes there will be a small child standing by to drain any cola remaining in the returned bottles. If a ton of cullet is used to make new glass, it eliminates the need for more than a ton of raw materials. In addition, the use of cullet in the batch lowers the temperature required for melting, cutting the cost of fuel for the furnaces. The major expenses in making commercial glass, after the cost of the materials, are for labor loyal to a tradition of strong unionism (among the original trade unions in Samuel Gompers's American Federation of Labor was the Glass Bottle Blowers Association, or GBBA) and for the natural gas fuel (in the fabrication of plastic, the killer expense is for the petroleum on which the material is based). Manufacturers, therefore, are eager to reach into the millions of tons of solid waste streaming from the seven thousand curbside recycling programs in the United States to retrieve as much glass as they can; one large maker of containers, Owens-Brockway, paid $300 million to purchase recyclable glass over a period of ten years. Once collected, the cullet must be separated according to color after nonglass contaminants have been removed.

Plastic, of course, is recycled for use in the manufacture of mostly secondary products. This process of taking used plastic containers and making, say, paintbrushes out of them is called "cascading." The product changes, but the life of the plastic goes on until it is discarded in the landfill. Cascading occurs with recycled glass also, when the waste is turned into glass pellets used for batch-making and other purposes. But for the most part, glass goes full circle, from container to cullet to container. Because of its inertness, glass does not have to be recycled for the sake of the environment. It is not readily dissolved, except by hydrofluoric acid, and it does not allow toxic leachates to escape from it. If crushed fine enough, glass can be returned to earth almost as it was in the beginning.

It is a never-ending challenge to find new uses for the salvaged cullet that does not go for another melt in the furnace. One sparkling (literally) fruit of that effort is "glas-phalt." By substituting crushed glass for the sand and crushed stone that goes into the making of conventional as-phalt, there is a substantial financial saving for municipal road departments. In addition, the mix offers durability and a measure of improved appearance for a road. It has been used on the streets of Brooklyn and Omaha, and on stretches of highway in a number of states. In South Carolina, not far from Aiken, there is a piece of glasphalt road that runs where azaleas and wild dogwoods grow, and sunlight strikes it in such a way as to forge a galaxy of glints, all too soft and fleeting to raise fears of tires being punctured.

And in southern California, a lawn mower dealer named Dan Dalager has dreams of restoring eroded beaches with crushed waste glass. "I was on vacation along the coast in the northern part of the state, in a town where they used to throw the garbage into the sea. Everything washed away except glass and other silicates, and when I saw all of it there, I thought, 'Why don't we use glass to build up our beaches?' " he said. "When crushed, it looks just like sand, feels like sand, and it's so beautiful in the sun. It glimmers and it's something to behold."

It is a valid concept. The beaches in the vicinity of his hometown, Encinitas, need an additional one hundred mil-lion tons of sand. It is a seemingly bottomless pit of erosion, capable of taking all of California's recycled excess glass for many years to come. At first there was skepticism, hard questioning: Would it be safe to walk with bare feet on the glass? What if there are contaminants in the glass? How much will it cost?

"Although one company offered to give two hundred thousand tons of ground glass for use on the beach, the project was killed for what they claimed were economic rea-sons," Dalager said. "It was all just more proof that government isn't here to get anything done, but to make sure nothing gets done wrong. And the best way to ensure that nothing gets done wrong is to see that it doesn't get done at all."

Had the plan gone through, the glass beach at Encinitas would have been the first in the country, but maybe not worldwide. During the Second World War, allied aircraft on bombing missions over Nazi Germany's chief submarine base in the port city of Kiel were reported to have destroyed a brewery in the region, and one resident of Kiel at the time has recalled that the broken beer bottles were further crushed and hauled off to the beaches of that Baltic port.

Dan Dalager has heard from a small Illinois community near Chicago, on the shore of Lake Michigan, concerning his plan. After hauling in as much as half a million cubic yards of sand over the past five years to fight erosion, officials of the town told Dalager that they have an interest in using crushed glass. But nothing came of that, and Dalager's proposal, while promising, quickly faded. It happens so often: to the man who reveals that he can make fuel to run an automobile by boiling the leaves of a certain tree in a remote part of India; to the chemist who announces he is on the verge of perfecting a new tomato that can be picked and sold ripe, and a peach that need not go to market fuzzless and as hard as a billiard ball. On learning of such wondrous achievements in the making, there is that first flush of voracious hope, and then the dulling realization that we'll probably never hear of them again. We usually don't.

A degree of resignation is to be detected when Dalager speaks now of glass on the beach. He can be faintly amused at the irony of it all, the fact that much of the erosion of Encinitas beaches can be traced back to past years when sand was hauled away from there to be used to make—what else?—glass.

Among the largest consumers of cullet are manufacturers of fiberglass. As much as 40 percent of the glass used to make the product has been recycled. Imitating in many ways a worm making silk, fiberglass-making machinery forces droplets of molten glass through apertures, then draws it long and thin by the centrifugal force of rotation. Cooling and solidifying, the fibers wad and the batting is marketed as insulation—long, itchy pillows stuffed between joists to defang the bite of winter. Fiberglass insulation is, for the most part, an American product for American consumption, being limited in that regard by foreign import restrictions and the preference in Europe for rock wool insulation. Also, transporting the cotton-candy-like material presents a problem. To be shipped abroad economically, it would have to be compressed to the point where the binders on the package are likely to fail.

More than a stronger binder, the glassmaking industry today needs a lighter container. A consumer may prefer the look of a product packaged in glass, but plastic inevitably wins the carrying-comfort competition. Those in the glass end of the container business are waiting, hoping for the breakthrough to a lighter glass, to the time when jars and bottles need not be double-bagged at the supermarket. Although rare, there are instances, however, when the heft of glass is an advantage, as Dr. Peter Vergano, associate professor of packaging science at Clemson University, reminded me. "When a customer pays three dollars for a couple of ounces of instant coffee," he said, "that person is going to feel less cheated if it's packaged in glass because it will weigh more."

Instant coffee aside, a lighter glass would lower production costs. "If we could make a bottle two thirds the thickness of what it is now, we could save enormous amounts of energy and manpower," said Dr. Arun K. Varshneya, professor of glass science and engineering at Alfred University. "We are working on it." He has calculated that by reducing the thickness of a glass bottle by 40 to 50 percent, the weight will be reduced by 30 to 35 percent. There has been some progress in that direction in past years, but not enough to mount a serious challenge to plastic.

It is not enough to simply make the walls of the glass bottle thinner. There must be a compensating increase in the strength of the glass. In the case of a 30 to 35 percent reduction in weight, the strength of the glass would have to be raised by 100 percent. It all comes down to the undying, seldom yielding, always nagging bugaboo of the glass business. "In addition to weight," Dr. Varshneya said, "glass is also at a disadvantage because of its lack of impact resistance. It breaks."

Glass breaks, and that is the insuperable truth.

And yet the material has an intrinsic compression-resisting strength greater than that of steel. It was once theorized in a Corning promotional brochure that a one-inch piece of high purity glass "suspended in mid-air could support 216 six-ton elephants." That may or may not raise a number of questions from someone who has opened a gift box to find a small glass figurine in shattered ruin.

Theoretically, yes, the fiber of optical glass could hold the tonnage, since the birthright strength of glass may be as high as three million pounds per square inch. But in reality, the elephants would be in for a fall. Strength can be characterized by a modulus of rupture: How much force can be applied to glass before it breaks? That's the strength of the material. The other crucial property is its toughness, and that is measured by how far the material bends before it breaks. Glass has the strength, but it takes little more than a scratch on the surface to sap off much of that, as one sees when a windowpane is cut to size. A flaw or imperfection in ordinary glass is like a malignancy. Just touching the material when it is being made can bring about potentially fatal surface imperfections. In the end, the strength of ordinary glass that has been cooled by annealing has been reduced to a maintenance level of a thousand pounds per square inch.

Determining tolerances is a crucial factor in the engineering of glass. At Corning, Dr. Augustus Filbert, then director of research development and engineering, was explaining that the heat in the small laboratory furnace registered 2,100°F. "What we're doing now is testing the elasticity of the glass, the stiffness," he said. "How stiff is glass?" He posed the question not in anticipation of an answer, but as a transition to what came next. "So we take a piece of glass like this, with these strings on it—you see the strings?—and we lower this specimen into the furnace. We're over two thousand degrees Fahrenheit in there, but the strings do not burn in half. That's because there's glass in them.

"We shake the string on one side, and we pick up the signal from the shaking on the other string. We're looking for the resonant frequency, where we receive the maximum amount of vibration. We plug that information into a formula, and the computer takes all the data, and with a push of just a few buttons we have just what we want to know. It is important that the margins of elasticity for each type of glass be known."

In general, glass is a material with almost perfect elastic properties. It can be stretched and bent, and it will revert to its natural shape, which is more than can be said for the steel rod that the circus strongman bends. Light a single lamp near a large window during a time of gale-force winds and watch the way the reflection is distorted as the glass bends. There is no permanent distortion, other than, of course, the shattering and lasting destruction that occurs when elasticity is pushed beyond its limits.

Glass cannot be broken by compressing it. Tension is the culprit. It is when the material's tensile strength is insufficient to hold a microcrack together that breakage occurs. Scientific measurements have revealed that a crack in glass can start to tear at a speed of one trillionth of an inch an hour, and then accelerate to as much as 3,300 miles an hour. The speed is determined by the amount of tensile stress at the point in the glass where the crack has stopped. If the stresses are removed, the crack will progress no further, but adding to the pull on the material is like stepping on the accelerator. It is when molten glass is subjected to rapid changes of temperature—especially from hot to cold—that it contorts and stretches, strengthens and becomes stressed. While thermal shock can increase the risk of breaking, it can also give considerable strength to the material provided the two extremes of temperature are applied evenly and the cooling is uniform. It is all part of the heavy paradox in the breakage factor of a material that can be very tough and yet, at the same time, as fragile as a saltine.

Tempering occurs when the heating and cooling are done with pinpoint timing and distribution. Cooled quickly, the two outer surfaces become the ice of glass; the inside, meanwhile, still hot, continues to contract, increasing tensile stress on the glass and therefore adding to the strength-giving surface compression until pressures of twenty thousand pounds per square inch and more can be sustained.

King Charles II, in seventeenth-century England, must have thought an alchemy of illusion was astir when Bavarian-born Prince Rupert (for whom Canada's Prince Rupert's Land is named) came before the royal court to demonstrate this erratic behavior of glass. He had let droplets of molten glass fall into a bath of water, allowing the exterior to cool before the interior, setting up strength-giving surface compression as well as tensile strength. When congealed, the glass, now called a Prince Rupert drop, takes the shape of a tadpole, and it is the tail that becomes the Achilles' heel. The prince drew gasps of amazement from members of the court when he smashed a maul down on the body of the piece and it remained intact. But when he snipped the tail, the entire drop gave way to the stresses, causing explosive, shattering fractures to carry through the glass, leaving only granular particles. Such a demonstration only reinforces the words of a noted glass engineer, Dr. Frank W. Preston: "All glass dies a violent death."

As far back as the turn of the century, Corning was searching for a way to make a glass that could withstand sudden changes of temperature and be more resistant to heat. The need was underscored when a freight train signalman's lantern failed during a storm. Cold rain had fallen on the warm lantern, and the glass shattered; hurtling through the dark, an express collided with the unseen freight train. The solution came with a new glass-making formula (Corning currently has hundreds of thousands of different formulas on file) in which borax was added to the raw materials.

In addition, World War I was then in the wings, and the time was close when Germany, a nation with a history of deep commitment to glass research, would no longer be a source of special, highly advanced glasses for the United States. Such glasses were essential for use in medical and other scientific fields, and so the quest for new achievements by Corning was given added purpose.

Improving on its heat-resistant and stronger glass for lantern globes and battery jars, the company went on to develop new borosilicate glasses, which expand only a third as much as regular glass when heated, and even less than most metals.

It was a breakthrough of profound importance not only to glassmakers, but to homemakers and scientists and a number of sectors of the national economy. Being both heat-resistant and immune to degradation by chemicals, borosilicate glass was ideal for use in laboratories. Automobile companies found a variety of uses for the glass, including the making of sealed-beam headlights. And as a pan in the oven, the material could incubate a meatloaf to perfection—glass, unlike shiny metal, absorbs radiant heat. It has been almost eighty-five years since the material was first introduced and given a name that quickly won recognition in kitchens throughout the land: Pyrex.

Much research continues to focus on making glass even more resistant to heat and expansion, a lighter glass with added strength. A process involving ion exchange has been developed and is now in limited use. Ion exchange is activated by placing glass that contains sodium into a liquid mixture containing potassium nitrate, and then heating the combination to just below the transformation point where the glass begins to soften. The sodium ions of the glass move into the melt of the chemical salts, and the potassium ions go to the glass. Because the new ions in the glass are larger than the departed ones, they take up more space, thereby putting the glass under additional strength-building compressive stress.

Cramming the larger irons into the glass also serves to close surface cracks. "It is like heavy rain falling on a cracked and arid desert floor," said Dr. William Lacourse, a glass scientist at Alfred University. "The moisture goes into the soil and the cracks close. But ion exchange takes a lot of time, and if we can cut down on that, the benefits could be widespread."

In that regard, Dr. Lacourse has found that bombarding the mixture with sound waves helps speed up the process, thereby cutting costs. Although ion exchange was in use as long ago as the 1960s, it has been restricted because of the expense. When atoms move, they go first class.

But there are those who are willing to pay the present costs. Among them are aircraft companies, and companies involved in the space program (the windows of the Apollo craft that carried astronauts to the moon were made of ion-exchanged glass), as well as makers of eyeglass lenses, medical equipment such as centrifuge tubes for use in the laboratory, and, moving to the battlefield, ampules to hold the antidote for nerve gas.

Since the introduction of Pyrex and ion-exchange glass, many other new families of specialty glasses and glass-ceramics have been developed, and some are tough enough to shape on a lathe or fashion into nuts and bolts. Others can withstand the heat on the skin of a spacecraft during reentry, and move in protective step with some forces of breakage. Here is Walter Cronkite remembering the first launch of the Saturn rocket: "They built these quarters for us, broadcast booths for each of the networks, and we had a two-story place with a big glass front. . . . The first unmanned Saturn went up . . . there was an earthquake . . . the acoustical tiles were falling around us . . . this big glass window was shaking and we're holding it with our hands. I put my hand up to . . . hold that window in place. . . . It was only afterwards that the glass people came rushing in and said, 'Don't ever do that again. The glass is made to vibrate. If you stop the vibration it'll break for sure.' "

But those are largely scientific exotica; they do not speak of the ubiquitous, low-cost soda-lime glass that makes up at least 75 percent of all the glass manufactured in the world today. The courtship of that glass by breakage will continue for many years to come. Rocks and balls will continue to come crashing through windowpanes, BB pellets will continue to take out streetlights, and shoppers will continue to mutter, "Aw, damn," when a jar slips from their hands and shatters on the unyielding floor of the supermarket.

There may even be more television commercials for Memorex tapes, in which the recorded voice of a vocalist manages to shatter a thin-stemmed glass goblet. In order to break the glass, the taped voice must be singing the same note—or frequency—that is heard when a finger is rubbed around the rim of the goblet. And then, by raising the decibel level of the voice, the glass *may* come apart. It is not easy to do, and the commercials have raised questions about truth in advertising. But then, the late and great Ella Fitzgerald was a party to this, and so there is reason to be trusting.

Breakage can be brutally saddening for a glassblower shaping a work of art, since it occurs most often in the final stages. The artist has been to the furnace a dozen times for new heat to keep the glass at a consistency somewhere between that of honey and a well-chewed Tootsie Roll. He has worked and reworked the piece until it is ready for the annealing lehr, or oven. In two weeks' time it will be in a gallery, softly backlighted and available for a considerable amount of money. Only one thing remains, and that is to release it from the blowpipe. The pipe is tapped to break the connection, but the large glass, meant to fall into the open and well-padded arms of an assistant, is off course. It is also, quite suddenly, all cullet.

Of breakage, William Morris, a resident of Washington state who has found a place in the top ranks of glass artists worldwide, has this to say: "The glassblowing process is very humbling, and I have always been appreciative of how much I am able to get away with."

Short of producing a flawless glass, the best hope for controlling breakage rests in finding a way to prevent a crack from accelerating to shattering speed. Why do some cracks expand for as long as ten years before the glass fails, and others for just a fraction of a second? Plain water is among the agents that can speed up the expansion rate of a crack by leaching certain components from the glass, which then form a corrosive layer on the surface. That can have a strength-sapping effect on the material.

On the other hand, water on glass can be a heavenly presence, as was the case with certain of the windows of an office building in Clearwater, Florida. It began in December 1996, with the first sighting of what resembled a brightly colored portrait of the Virgin Mary on the glass. The devout and curious soon came by the tens of thousands, and droning intonations of the Hail Mary carried through the air. For weeks people came to see the "Glass Madonna," and few of them, understandably, were willing to accept the plausible theory that the portrait was caused by a chemical reaction between water and glass. Being double-pane coated glass, the discoloration developed probably when moisture got between the panes and eventually eroded the metallic coating. The optical effect resembles an oil film on water. As for the figure of the Virgin Mary, the only explanation, short of divine intervention, is that shadows from a nearby tree fall on the glass in such a way that the shading acts to define the corrosion.

There is a sweetness to both the mystery and vulnerability of glass. Fortunately, there's nothing that can be done about the former. As for the latter, the tinklings and explosions of breaking glass will continue to be heard for now. It may be best if it's that way forever, for to deny glass its fragility would be to chill the romance of the material. And pity the loss of a morality lesson, because then, you understand, people in glass houses *could* throw stones.

Chapter Four

Through a Glass Flatly

There are vistas to behold in this world, and it is glass, as often than not, that brokers the viewing.

We ride in a tour bus, and the oversized windows and windshield drop the passing countryside into our laps. A storm is up, and we are housebound in a place glass-fronted to the sea, and the spectacular rage is at the windows, wanting in, filling us not so much with fear as with exhilaration. And there is the kitchen window of a house through which we see the garden and the great stirring of mayflies in the spring. It is flat glass in those windows, the glass the world knows best, the meat and potatoes of a global glass industry with sales of more than $40 billion a year.

The process of making flat glass by first blowing a cylinder by hand, splitting the piece in two parts, and then ironing both flat was still in heavy use in the early years of the twentieth century. The first movement toward a mechanical system involved the use of compressed air to form the cylinder; it was always the blowing by human breath that challenged inventors, for it was a sweat-soaked and labor-intensive method of giving shape to molten glass, a nightmare of inefficiency little changed since before the birth of Christ.

John Lubbers answered the challenge in 1896, using his good mind and experience as a glass worker in Pittsburgh to create a machine that automatically gathers molten glass and lifts it up in the form of a cylinder. A vertically moving blowpipe with a flange around the bottom is lowered into the molten batch, and then, as the pipe is raised, with the glass clinging to the edges of the flange, compressed air is blown in through the top, giving a bubble shape to the glass. The air pressure must be steady and the speed of the lift without variation, and when that is done, the rise of the glass—cooling as it goes—is true and tall. It remains then to lower the cylinder to a horizontal position so that it can be cut into sections that are slit, heated, and flattened. It got so that the system could pull up a cylinder as tall as fifty feet and with a diameter of almost three feet, although lowering such a large and fragile piece onto its side was not without considerable risk. It had to be done by cradling the piece in a sling and then inching it down by cable.

The invention served to make flat glass more readily available to more people, at a lower cost, but problems regarding poor quality remained. There were surface imperfections caused by the ironing of the rounded cylinder halves. In addition, the pieces could not be made perfectly flat. For these reasons, the view through old glass is often distorted and the glass itself bowed. Nevertheless, the mechanization resulted in a doubling of the amount of flat glass being made, as well as a severe cutback in the workforce.

It was a heady time for persons of vision, those very early years of the new century. Henry Ford introduced his Model T, and Roald Amundsen blazed a seafarer's trail through the Northwest Passage. And there were those with an interest in glass who diagrammed their thoughts about easier ways to make the material in flat sheets, and they filed for patents, built prototypes of the new machinery, arranged financing, and went into production.

They showed the world how to make flat glass by rolling: The molten glass is herded through a gate, and then, as a sheet, moved laterally through two water-cooled rollers and on into the annealing lehr. The process works as a flattener, but the glass is of poor transparency and therefore of limited use. (It is sometimes still seen in the public toilets at national parks.) And they devised a method for molten glass to be drawn up in sheets. Irving Colburn, a Pennsylvanian, and a Belgian engineer named Emile Fourcault worked separately, but simultaneously, to develop the process. In 1912 Colburn's patent was obtained by the Toledo Glass Company, forerunner of today's Libby-Owens-Ford Glass Company, still headquartered in Toledo.

In the early forms of drawing glass, a *débiteuse,* a ceramic box with a slit in it, was floated on the molten glass, and when slight pressure was applied, the box pushed down to allow the glass to rise through the slit, where it affixed itself to a piece of metal bait. Caught then in the grip of rollers, the glass was drawn up in a sheet more than twenty-five feet high. The glass was cooled as it was drawn up, and in that way it became solid enough to resist most surface damage from contact with the rollers. Libby-Owens made improvements to the process, allowing the molten glass to be drawn directly from the surface using the metal lure, but not the box with a slit. The glass is drawn quickly and grasped by side rollers to keep it from narrowing before it gets to the annealing lehr.

A still more advanced system of making flat glass would be developed by the Pittsburgh Plate Glass Co., now PPG Industries, in the late 1920s, but it too was not without its problems. The systems were flawed because of deterioration of certain parts, and contacts between the glass and ceramics in the furnace. Also, the thickness of the glass was not always constant, and therefore the images seen through it were often distorted. It was classified as simple sheet glass; to reach plate glass status, it had to undergo extensive grinding and polishing. Nevertheless, the making of flat glass without first having to blow a cylinder represented a major leap forward for the industry. The process is still in use, although it has been pushed into near obscurity by the widespread use of a glassmaking system introduced just thirty years ago.

Corning and Pittsburgh both missed out on this one. Rather, it was England's giant glass maker, Pilkington Brothers Ltd., that scored the beat. The company revealed a method of making flat glass by simply floating the molten glass on a bath of liquid that will not give way to pressure while gravity flattens the glass against it. In concept, it would be a simple procedure, almost like laying cream on coffee. Almost, but for this: The chemistry is complex, in that the liquid bath on which the glass floats must be a molten metal that will not chemically react with glass, that will serve as a heat exchanger, and will be able to tolerate a temperature of 1,850°F without vaporizing and 1,110° without hardening. There is only one metal capable of performing in that manner: tin.

The float occurs in a heated chamber under strict atmospheric controls to keep the tin from oxidizing. There is a continuous flow of molten glass onto a two-inch-thick bath of liquid tin. The glass is about three feet wide at the mouth, and when it settles on the tin and starts its 180-foot journey through the chamber, it spreads out, and tractive pressure is applied to maintain the new thinness—a quarter of an inch—of the glass. It is not enough for the tin, with its high density, just to support the glass. It has to function as a conveyor belt, and it must pass along heat to the underside of the glass while sealing out moisture. Tin can do all of that. Other companies followed Pilkington with their variations on the process, but the fundamentals remain unchanged.

At a float plant in Rossford, Ohio, in the forlorn shadows of Toledo, the natural-gas-fired furnaces consume fifty thousand to sixty thousand pounds of cullet-heavy batch every hour, and from that a steady ribbon of molten glass emerges to be transported by rollers to the bath of more than 150 tons of molten tin. Once atop the tin, the glass is exposed to a stream of hot air from above, and that, together with the contact with molten tin on the underside, acts like a polisher on both sides of the glass. Out of the annealing lehr, it is flat glass of consistent thickness and brilliant transparency. In addition to built-in quality control, the float method embraces speed—it is faster than all the other ways of making quality flat glass.

Laser beams are reflected through the glass to reveal any defects, and then the glass is cut by being scored with carbide cutting wheels before being snapped off mechanically. After a protective dusting with a powder, the pieces are packed for shipping. There is less breakage in the float process of glass production, but, as an official at the Rossford plant said, "It's still more than we like." That is not to say, of course, that they neglect to take precautions; indeed, a warehouse there where thirty thousand to forty thousand tons sit in storage has been designated as a tornado shelter.

As much as 95 percent of the flat glass produced worldwide today is made by floating it on tin. There are thirty-seven plants in the United States in which that process is used as the most economical and efficient way to provide the signature form of the material, flat glass of good quality. Much of it serves as the base material for the making of windshields and windows for vehicles; at Rossford, almost all of the production is for automotive use. Another plant in

Meadville, Pennsylvania, a massive six-hundred-thousand-square-foot facility of PPG Industries, produces a million square feet of flat glass a day—enough for the windshields and windows of twenty thousand cars.

We make mirrors and furniture of it, and as a construction material it dominates our urban landscapes, a role in which it takes on a life of its own. When the glass of a building dies, so does the building. (Sighted from a passing train, is it not the black holes of broken windows in an abandoned trackside factory that best signify the heavy despair of that place?) Based on a study done in the late 1970s, the "broken windows" theory has been incorporated into the crime-fighting strategies of some major police departments, including that of New York City. The theory proposes that if a window in a building is broken and goes unfixed, the assumption will be that no one cares, and therefore it's okay to break the other windows in the building—an attitude that then fosters more crime in the neighborhood. Thus window breaking is one of the so-called quality of life crimes that New York police have concentrated on in recent years, resulting in reduced crime overall.

Flat glass is often chemically altered and combined with other materials. So it is that it becomes a windshield of coated glass encasing plastic. With each passing year and auto model, the windshield becomes not only larger, but more technologically complex. There are different methods of curving the windshield glass, one of which is to have the still-hot glass slide into a peripheral ring and then sag to the desired shape. But with auto makers now wanting more acute curvature in the glass, attention is being paid to a possible pressing process, using a double-sided mold of male-and-female configuration.

Of course, we already have windshields that reflect solar heat in addition to melting ice and zapping fog, and windshields engineered to serve as the antenna for the radio. And, no doubt, there is more to come. The rear window may serve as an antenna for the cellular phone, and for a system to give readouts on global positioning. Data from a satellite is projected onto a video screen inside the car, giving the driver his location and directions to his destination. With some systems, a recorded voice offers information, including alerts on when and where to turn. At present, the positioning system is in limited use in the United States, available on some high-priced cars as a $2,500–$3,000 option.

There is now windshield glass that automatically sets the wipers in action at the first splash of rain. And HUD—the "head-up display"—may be a windshield feature ahead of its time. It was developed in the late 1980s, nothing to rank with front-wheel drive, but a small jewel of inventiveness nonetheless. A six-inch-square screen is laminated into the windshield; it is a holograph, and invisible. The dashboard readings are projected onto the holograph, and that in turn makes them visible to the driver in a spot that appears to be about ten feet in front of the windshield, out beyond the end of the hood and slightly to the side so as not to block the head-on view. It eliminates the need for bending and craning to get a read on the car's speed or the amount of gas in the tank. Head-up display systems have been used in aircraft for nearly fifty years, but their use in automobiles has been limited to two or three models.

Makers of flat glass devote much of their research and development to serving the needs of the automotive industry. "Once, the drive for advances was primarily focused on glass for buildings, but then it shifted to the automobile," said Dr. Roger Scriven, director of glass research for PPG. "We're doing things today that were impossible five or six years ago." The windshields that ward off solar heat and the searing discomfort of a baked four-door allow the use of smaller, less expensive air-conditioning systems in vehicles. Research into making windshields of different colors, to blend with the exterior color of the cars, is well along, and so are windshields treated to repel rain in such a way as to do much of the work of the wiper. All of that is done with coatings on the glass, sometimes of multilayered stacks of different materials.

As with the automobile, flat glass made by the float process is the basic material for the manufacture of windshields ("transparencies," glassmakers prefer to call them) for many commercial airliners. The use of glass in planes dates back to the mid-1920s, when plate glass was fashioned as roll-up windows for the Ford Trimotor. By the 1950s, when large-scale commercial air travel was beginning to take off, the glass being used was laminated and heated. It was also made strong enough to withstand the impact of birds—not swallows or others small and fluttery, but geese and ducks and other heavyweights that can deliver artillerylike damage to an untreated airplane windshield.

There are glasses in aircraft today that are more than five times stronger than the ordinary material. This is done by adding lithium to the batch for making float plate glass, and then bathing the formed glass in molten sodium nitrate, thereby setting up that magical ion exchange: The bulkier sodium ions crowd out their smaller lithium counterparts, thus strengthening the glass through compression.

In the old Boeing 707 aircraft, the glass windshield in the cockpit had a life span of just one year, and even that was extraordinary considering the heavy weather, abrasion, and the threat of impact are frequently in attendance on a flight. Today, with lithium-strengthened glass, that life span has been extended fivefold. In addition to being stronger, the glass is lighter, thereby depriving plastic of its wide edge in weight. Polycarbonates and acrylics continue to be used for the transparencies in some aircraft, mostly military, but glass is the leading choice by far. Indeed, all windows and windshields in all Boeing airliners today are made of glass, as are those in the huge C–130 cargo plane with its twenty-three forward-facing windows.

It is in the windshield of an airplane, as in the transparent skin of a skyscraper penthouse, that glass most forcefully exudes a sense of strength and protective power. And that is especially true at night, when, flying over the ocean, the black sky and all its demons of atmospheric uncertainties are rushing at you, sometimes with rain and lightning. As the country singer plaintively deems himself just "a bug on the windshield of life," so does the pilot see his life (and the safety of it) reflected in the glass before him.

Early motorists and fliers alike must have given thought to how nice it would be if the windshield could automatically turn dark in sunlight and regain its clarity when the light fades. The science for doing that is at least thirty-five years old, dating back to the mid-1960s, when Corning unveiled its photochromic glass.

But even then the process wasn't completely new, for nature has been doing it for as long as anyone knows. It happens in the desert, where you might find a purple bottle long abandoned, a relic from a time when, say, the S. S. Kresge store sold Jew's harps for a dime each. The bottle, once clear, has become solarized by its decades-long exposure to ultraviolet rays. When the energy of the rays allows an electron to escape from the colorless manganese, the valence of the metallic element rises, causing it to oxidize and turn purple.

Similarly, up on Boston's Beacon Hill there are old houses with chameleonlike windows that show pink in the sunlight at certain times, and lavender and heliotrope at other times. Actually, the panes are purple, having turned that way after a few centuries in the rarefied light that falls on that citadel of manners. Some believe that the glass, imported from England, was purple from the start. But why, then, is it white and clear around the edges, where it is shielded from light by putty?

The thought of a windshield that could react to sunlight like that, and do it right away rather than over a span of years, was compelling, and all the more so because Corning scientists had succeeded in achieving rapid solarization in the laboratory. They did it by adding small amounts of silver chloride to the premelt batch, thus leaving the finished glass laced with billions of small crystals of silver chloride. With sunlight playing on the treated glass, electrons are borrowed from the chlorine atoms to reduce some of the silver to the metallic state. It is then that the glass darkens, turned gray by the metallic silver particles. Remove the sunlight and the electrons return to the nearby chlorines, allowing the colored glass to become clear once again. And while it takes years to color untreated glass, the speed of photochromic reaction is defined more or less by the time it takes the glass to move in or out of sunlight.

A photochromic windshield seems a natural option on a new car, except that the glass is expensive. And something else: It doesn't take long for the glass to recapture its clarity, but it's not instantaneous; therefore, what's a driver to do when the car goes directly from sunlight into the darkness of a tunnel? It was decided that photochromic glass was best excluded as a material for the windshield.

But used for eyewear, the glass that darkens on its own has been triumphant.

"In terms of profits, photochromic glass as lenses for spectacles was a huge success for Corning," one of the company scientists told me. "They made a lot of money, a killing. It was an item that was unique, and they could charge a lot of money for it, although the production cost wasn't that great. So you have to ask yourself, how come all the companies aren't into photochromic glass? It's because it's so difficult to make. Over a period of years you have to experiment in making such glasses by doing a little bit of this and a little bit of that, and controlling this and that. Only with expertise of control and precision could photochromic glass have been made."

Indeed, the Corning complex and the people who work there, among the company's more than forty thousand employees worldwide, are a study in control and precision. Men and women with advanced degrees from quality universities are in residence in the small offices along long corridors, and they are scientists who seem scantly possessed of the bonhomous spirit. A raised voice is seldom heard, and God knows the last time there was any hanky-panky at an office party there. There is a strong loyalty among them to the tradition of research at Corning—laserlike research that has burned through the walls of imponderables to give the company title to most of the major advances in glass technology over the past century. The company held almost all of the patents for the use of glass in the making of a television set, and it licensed the rights to the technology to the world. Corning developed a more efficient method of making the funnel of a television tube—the bulky, martini-glass-shaped piece that sits between the front display panel and the rear neck—by centrifuging, or spinning. Previously, glass was pressed into a mold; now, a glob of molten glass is dropped into a centrifugal mold device and whirled into shape.

From Edison's light bulb to the cathode-ray tube, Corning has performed brilliantly in packaging electronic devices in glass. Of course, the transistor has replaced the vacuum tube as a means of amplifying current, but the cathode-ray tube, mass-produced by Corning during World War II for use in radar detection systems, remains an integral component in the television set as well as in more sophisticated items in use for national defense. Researchers tirelessly explored new ideas: fiber optics, glasses for liquid crystal display, machinable glass-ceramics. There was a sense of self-perpetuation, that something new would always come along. It even seemed sometimes that the company became bored with a successful product because of anticipation over the next breakthrough in research and development. Glass is so versatile, so adaptable, that it fills a scientist's mind with the rapture of discovery, be it the small dosimeter locket of glass treated to measure the amount of dangerous radiation the wearer may have received, or a method for casting a single piece of flawless optical glass weighing more than a ton. Triumphs in the glass labs do not come easily, but they are there for the finding.

But now at Corning and some major glassworks, there is pressure to rush the finding. It is possible to detect a touch of the bean-counter virus at work. Taking note of that, a scientist of long service at Corning said that depending on whom you are talking to, "research in the company labs now tends to be anywhere from two to ten years away from reality. The scientist says ten years, research management says five, and the business office says two."

Despite the thousands of years of its use, glass remains a material with hoards of secrets. But if that were to end now, with no more disclosures, the world would still have derived what may be the best of all that glass has to offer, and that is its defining role in establishing a visual relationship between openness and protective enclosure. Only glass can set up a face-off between the two, and it does that most effectively in its plate form.

It was in 1989 that Donald Trump was in his office on the twenty-sixth floor of his Trump Tower, in New York

City, signing checks as we talked about his real estate holdings. "And that," he said, being courteous and very likable, and gasifying the air with his ego, all at the same time, "is the Plaza Hotel. I bought that." He made an imprecise gesture toward a window behind him, and there was the hotel, blocks away, but drawn by the glass to hang before its new owner like a snapshot on the refrigerator door, a trophy in its case.

That same year I went to the top of the Empire State Building with the actress Fay Wray, last seen at that height while in the gentle grip of King Kong, the great ape of filmdom (suffer the truth and know that the film was shot in Hollywood, using a mock-up of the skyscraper no taller than a cabin in the woods). Eighty-one years old then and an enchantress still, Miss Wray was visiting the tower for the first time since 1934, the year following the movie's release. Now she was standing at the top, on the observation deck, not in the shadow of Kong, but in a bath of sunlight falling through glass. Panels of flat glass were in place around the deck to keep the 1,250-foot-deep abyss at bay, inaccessible to suicidal jumpers. The panels also served to make the breath-sucking height of the place less fearful by far.

Trump may have the Plaza, but Fay Wray was laying claim to this one. She spoke above a rush of wind: "This building belongs to me!"

In the world of urban towers, flat glass is a ubiquitous fixture of both design and function. As the skin of many soaring post–World War II structures, closed off against the air and noise of the city, the material has done away with all but a scant sense of openness. For that, it has drawn the scorn of those who are enchanted by the coo of a pigeon on the ledge of an open window, and those who find no virtue whatsoever in being hermetically sealed away. Yet others find only pleasure in their secure encampment behind a wall of glass on, say, the forty-second level of an office building—an enclosure that sweeps them grandly and safely through space.

As a material for construction, flat glass was in use long before Pilkington introduced the float method of manufacture. Ludwig Mies van der Rohe, the late German-American architect, began designing buildings sheathed in glass as early as 1929. At first glass was used as a finish for outside walls, being attached with an adhesive or mortar. It wasn't until engineers devised a method where each floor of a highrise building could be individually supported on the steel spine of the building, thereby relieving exterior walls of support responsibility, that glass came into wider use. Then it could be hung on the frame like a shining curtain. Following World War II, glass buildings rose in large numbers, and among them was the Seagram Building in New York City, a thirty-seven-story masterpiece in glass and bronze designed by Mies van der Rohe and his architect collaborator, Philip Johnson. With that construction, and from then on, plate glass was the material of choice for the exteriors of towers.

Cities began to look spiky and Simonized, places with sparkle and sealed thrusts for height. Eight years after completion of the Seagram Building reflective flat glass came into use, and that had some high-rise centers aswirl in images. Colored reflective glass has succeeded in creating pleasant moods for some buildings, but it has also made others look like towering Lava lamps.

Whatever the aesthetic value of the glass, it is more now than simply cladding for the building. With the use of low-emissivity glass, it is now possible to control the interior temperature of the building while conserving large amounts of energy. It works to give the structure all the properties of a thermos bottle, keeping it cool inside when it's hot out, and vice versa.

Low-E glass, as it is called, is float-made flat glass with a coating of metal oxide, either tin or silver. If tin, the oxide is applied during the float procedure, as the glass is forming. That is called a hard coat. With the soft coat of silver, the application is made in a sterile atmosphere, using a vacuum process with the chemicals of the coating layered on the finished sheets of glass. Hard or soft, the coating works to repel the sun's heat and ultraviolet rays while at the same time allowing visible light to enter the building. In winter, the glass reflects the inner heat of the building, preventing it from escaping. Whatever the season, the savings in fuel costs are considerable.

"The use of low-E-type glasses for windows can amount to a savings of energy in the United States equal to that being supplied by the Prudhoe Bay oilfields in Alaska," said Roy G. Gordon, speaking at his home in Cambridge, Massachusetts. He is a professor of chemistry at Harvard, and he began working on his invention of hard-coat low-E glass in 1974. "It wasn't until 1988 that we had a commercial product," he said.

Low-E glass is double-glazed when installed, and in some cases argon gas, a good insulator, is injected between the two sheets. By itself, the material is five times more efficient than ordinary plate glass, in that it tends to differentiate between the three offerings of the sun's energy: visible light, ultraviolet, and heat-producing infrared. While reflecting the latter two, it allows as much as 95 percent of the visible light to pass through. Even so, low-E glass is usually used in conjunction with another solar-control panel, this one a glass often containing iron. And so it is that the new buildings of today's industrial parks—those expanses on the edge of town where once the circus was in business for three days each year—are often green, as that is the color that iron lends to the glass installed as the outermost piece, open to public view.

Bronze and gold and bright silver are other popular colors of the new glass, each derived by adding different metals to the melting batch. There is cause to wonder here if buildings, like kitchen appliances, may soon be in for periods of color chic. As the avocado-hued refrigerator was once the epitome of kitchen smartness, will the yellow (add cadmium sulfide) mid-rise have its brief day in the sun?

The time may even come when a building clad all in glass can be made to change color with the press of a switch. Again, it is the relationship of glass to electricity that has enticed scientists to envision startling new applications for the material. They have sent electricity through circuits placed in glass to keep windows in large freezers free of frost, and they have done the same to keep the window in the oven from steaming up.

That application of electricity is now being used to switch between clear and frosted glass. It is a system aimed at offering privacy when needed: the secretary's cubicle in an office building is partitioned off from other cubicles with clear flat glass. When privacy is desired, a switch is flipped and the glass becomes opaque. The flat glass used is laminated, with a polyester film in between the two pieces. The film itself is in two pieces, and between those is a matrix of liquid crystals. When an electric field is introduced in the film, the crystals align with that; when the electricity is discontinued, the crystals revert to random siting.

Actually, then, it is when the electricity is turned *off* that the glass becomes opaque, as the orderly grouped crystals block the light from passing through. In the opposite, inert state, the crystals are scattered like buckshot, with no mass to repel the light. The process involved here is not much different from that in a digital display, like that on a pocket calculator.

"There is a lot of interest in the new glass, but it takes time for that to be reflected in sales," a spokesman for the Minnesota Mining & Manufacturing Co. (3M), the maker of the polyester current-bearing film, said. "At the present time, the glass is in a stage of evaluation." It is a glass thought to have many possible applications for the transportation industry: farewell to the hard-to-pull little shade on the windows of airliners, and good riddance, too, to the gawkers who peer into the rear windows of ambulances. Use of the glass in the home could range from skylights to front-door panels. Architects no doubt see privacy glass as having a role in design, but the cost is as much as $100 a square foot when used architecturally.

With the basis of such technology applied to the exterior glass of a building, would it not be possible to have a structure with a face for all seasons, a leafy brown in the fall, an icy mint for summer? It may come about, although engineers are not yet prepared to transform, say, the John Hancock Tower in Boston into a pillar of green for St. Patrick's Day. But a London architect named Mike Davies can envision the day when "New York could perform a symphony of color as the glass in the tall buildings is made to change colors instantly."

Davies is a member of the Richard Rogers Partnership, designers of the new Lloyd's of London building. In that edifice Davies and his colleagues had opportunities to test their imaginations with applications of special glasses. The core of the building is a glass atrium, whose 16-story facade is fashioned of 14,350 square yards of glass. Into that glass were rolled thousands of prisms, to forge diamonds out of the sunlight and add sparkle to that towering house of indemnities.

As it must be with glass, sunlight figures prominently in the use of the material as a commercial product. And that is why I was standing outside of a mock-up of a two-story structure, pushing the buttons on a remote control to make the full-scale model slowly revolve, once around to get a visual reading on the sun's exposure to all the windows. It was summer, and a time of day when the sun was high, dropping a galaxy of light and warmth on Harmar Township in Pennsylvania. It is there that PPG has its Glass Technology Center, including the revolving house. "An architect can come here and see how the windows of something in design will look at any time of day, facing any direction," said Patricia Corsi, one of the 270 PPG people at the center. "We have a variety of glasses in the windows of the mock-up, and the architect can get an idea of how each will appear in different light. Some architects will even come here at night to see how the building and windows look then."

The house with the remote can sometimes work wonders, being just the thing to trip an architect's indecision and have it fall in favor of a PPG glass—an order, maybe, for enough glass to cover sixty stories.

Six stories, more likely. The time has come, some say, to mark the passing of skyscraper construction in most developed nations of the West, even though as recently as the late 1980s, I found, architects, developers, and disciples of computer-driven engineering were on a mission to build more, and build them higher and higher, to raise glass and steel and stone to record heights. It wasn't enough that the skyscraper had become the logo for urban development (in the large cities of America, anyway) and that it had bridged the twentieth century with its indestructible, prodigious presence.

It is reasoned now that the superhigh city tower is for nations intent on making a show of their economic coming-of-age. Such construction—some with setbacks, so that the plate glass sheathing rises like a terraced glacier—is heavily under way in the Far East and in Southeast Asia. Even if New York or Chicago, London or Toronto never again has a skyscraper raised within its boundaries, the dominance of glass in those urban landscapes will endure.

Much of the work at the PPG research facility is aimed at developing processes that make a simple soda-lime flat glass even more complex and magical than the highly popular (almost half of the residential window sales in the United States) low-E coated glasses. The same is true for research facilities at the partnership of Pilkington and Libby-Owens-Ford, and at Japan's Asahi. It is all research on the use of flat glass as a substrate for the dance of electrons and liquid crystals and protons—plain glass made exotic by the synergy of its pairing with metals and plastics.

PART II

GLASS AS SCIENCE: A MIRROR IMAGE

A man that looks on glass
On it may stay his eye;
Or if he pleaseth, through it pass,
And then the heavens espy.

—GEORGE HERBERT, 1633

Chapter Five

Into the Age of Wizardry

Nature gave glass to the world, and it took the world more than three thousand years to look that gift horse in the mouth.

All through the time of the use of the material by the ancient Egyptians and later by the Romans, through the Renaissance and the golden age of Venetian glass, then of service elsewhere in Europe and the New World, and on and on to the present—for all of that time, except for the past seventy-five years, glass remained a chemical anomaly. Scientific research on the material's chemical structure was scant and empirical. Through trial and error, glass gave what was asked of it, and it was that tractability that held scientific inquiry at bay.

There is still no clear understanding of the riotous molecular structure of glass, but scientists are now clawing to get at the atoms and make some sense of it all. "It is an industry, this effort to try and find out what goes on with glass," said Quentin Williams, an associate professor at the University of California–Santa Cruz, and one of those who are seeking answers. "We now have a decent understanding of glass, but not a terrific one. But this is glass, and if I didn't know that glass existed, I would never believe such a thing *could* exist. A solid liquid? It just doesn't make sense."

In this time of new and startling technologies, however, glass is making a great deal of sense. This extends far beyond the manipulation by blowpipe of a glob of molten glass, or, for that matter, the hanging of a glass curtain on an office building. This glass of the new sciences carries our telephone conversations, takes the heat of reentry as tiles on space vehicles, stores mountains of information as a disk, brings into sight for the first time the far reaches of the universe, and digests our nuclear wastes for permanent and harmless disposal.

Glass is doing all that and more, and still there is nothing to show that the utility of the material is even close to being exhausted. Before that can be done, scientists will have to get a clear reading on its lack of crystal structure, because that lack enables glass to be shaped—bent and stretched and subjected to all the other distortions of malleability. As far as anyone knows, no solid has a structural malady as severe as that of glass; even so, there are new findings to dispel the belief that the molecular disorder is complete.

"If I could shrink myself to where I could sit on a silicon atom in glass," Dr. Williams said, "what kind of view do you think I would get from there? Well, right next door I would see four oxygen atoms. That's order, but only as far as one angstrom [one thirty-billionth of a foot]. Beyond that . . . ? Now, if it was an atom in a crystal that I was sitting on, I would see everything in order, way off into infinity."

Rather than being pervasive then, the randomness of molecules in glass seems to be replaced in very small part by orderly grouping. Envision an inexperienced tile setter redoing the kitchen floor, getting the first row of tiles in perfect alignment, but moving off course in the next row, and then having the deviation expand until, at the end, the floor is in tiled chaos. But there is order in that first row, as it is with some silicon-oxygen linkages in melting glass.

The findings are based on research using X rays and the scatter patterns of neutrons beamed at the glass. It involves probing down to the bedrock depths from which the mystery of glass springs. Both theoretical and experimental research are at work here, with men and women using supercomputers and scanning microscopes and other exotic instruments to probe the elusive atom.

So far it has all resulted in, if not full disclosure, at least a better understanding of glass. As one Corning researcher said, "Less than one in ten projects involving new uses for glass succeeds. Now at least we have a chance of knowing why things go wrong."

Still, glass adapts easily to design and function, and so questions have been raised concerning the need to know the implications of the presence of some order among the molecules. Some say that to sanitize glass, to dispel its intimations of alchemy, would serve no purpose. Do what you will with glass, they say, but spare it the loss of its mystery. There is merit in such a reaction. Go sometime up into the Scottish hills overlooking Loch Ness, preferably when the heather is in bloom, and look down on the dark waters while giving thought to that most enduring of all legends about monsters. You will feel a need then for the mystery of it all, for the taste of the delicious unknown at a time when the moon has been laid bare, when the once forbidden poles are visited regularly, when magicians appear on television to reveal to the world how the pretty assistant survives intact after being sawed in half. It's all enough to make you crave naïveté.

Ah, but the advance of technology does not pause for reveries, and neither do glassmakers.

In Mainz, Germany, near Frankfurt, the Schott Glaswerke pushes down to the Rhine River, and from that complex have come many major advances in glass. More than a century ago Otto Schott was experimenting with different melt compositions in order to make a nearly flawless glass suitable for optical use. He was one of the first to vary density and refraction in glass by adding ingredients such as phosphoric and boric acids. Schott had a Pyrex-type glass before Corning had Pyrex, but, as has been the jinx of German glassmakers through much of history, wartime embargoes prevented the company from introducing its new product in foreign lands.

Today at Schott Glasswerke, glass is being engineered with evangelical zeal. There, the application of glass to science is as seamless as the procedure can be, and the results are often astonishing, especially in the field of optical glasses. Computer-based controls are very much in force at Schott, and it seems that the search for glassified molecular order is of little interest there. "With all the molecules and all the possibilities of their movement and interconnection and so on, even the largest computers are too small for the study of the structure of glass," said Dr. Peter Krause, Schott's chief of research. "We do not do such studies here, although it is a good field for the universities."

Glass is suited for the new technologies in a way that few, if any, other materials can be; because of that, the work being done at major glassworks today is, in some respects, a primer for the advances of science. The space shuttle *Columbia* was scheduled for a mission to test, among other things, the mating behavior of fish while in orbit. It was essential that the water of the fish tank remain clean. That was done by using a highly porous glass membrane that was stretched out as a hollow fiber, cut into tiny rings and placed in the fish tank aboard *Columbia.* The glass was an instant attraction for the microorganisms that feed on the impurities of the water. There they colonized and assembled their legions for war on all the licentious unlovelies that make fish tank water go green and slimy. When *Columbia* returned to earth, the water was clean and the fish population had increased by eleven. The glass used was made by sintering, whereby glass ground into powder is heated to a point just short of when complete fusion occurs. The powder, which was mixed with a binding agent, is densified, but porous, with the pores either connected or isolated. Such glass can be used for a variety of filtering procedures.

Porous glasses that are strengthened and resistant to thermal expansion are likely to be put in use as a new generation of filters for use in engines and medical laboratory work, including separation procedures. Glass opens the filtering process to visual monitoring, and, being resistant to corrosion, it can handle the transit of acids and gases.

It is little known, I am sure, but there are people in this world who spend a lot of their time worrying about rust. "I hate rust," a chemical engineer once told me with the same resolve that Louis Pasteur probably used when he told someone that he hated curdled milk. The engineer said he hated the color of rust, noting that it appears on metal like "those aging spots on the hands of old people."

Rust is more than a stain on the wrought-iron furniture in the garden. It is a costly scourge that corrodes metal in a variety of ways. Rust-resistant alloys are the answer, of course, and there are scientists at work today in that field. Such alloys under development must be tested in effluents of the acids that attack ordinary metals, and it is at that point that glass is called into service. In corrosion laboratories around the world, the acids—nitric, sulfuric, and others—and various additional substances with a taste for metal are in glass vessels; and there the experiments take place, for no matter how the alloys fare, the glass will remain without blemish. There are some metals that would do the same, but the visual advantage would not be there, and the costs, the stratospheric preciousness of, say, palladium . . . well, blessedly, there's glass.

Imagine, then, the use of glass in a plant that *manufactures* concentrated nitric acid. Schott, together with a Norwegian company, opened such a facility in England, using a special borosilicate glass for pipes—some nearly three feet in diameter—and vats, and many of the other components of a system designed to handle twenty-five thousand tons of nitric acid a year. It may be the only chemical plant in the world where a visitor can be made to feel like a voyeur, since the chemicals are sloshing around for all to see. For the first time, a plant charged with such heavy, metal-etching duty has a sense of the fanciful about it.

If it did nothing other than resist corrosion, glass would be a valuable material. In Germany, at Munich's new Franz Joseph Strauss Airport, there is a subterranean passageway of glass representing the highest technology yet known for ridding a facility such as a major airport of caustic materials. When a jumbo airliner is washed down, what falls off is a mixture of explosive, metal-eating agents, such as kerosene, paint residues, oil, and water scummy with detergents strong enough to cut through heavy grease. The system of glass drainage pipes is under the three maintenance hangars, and it extends for more than three miles to connect the facility with a treatment plant for the waste. The glass, another Schott borosilicate, has low thermal expansion, an important feature in the event that the petroleum waste catches fire in the pipe. Being nonporous and transparent, the glass is heaven-sent for this use.

Still, it has taken many years for the material to start to fulfill its promise, to reach the point where, more often than ever now, there is something for the research and development scientist among the tens of thousands of workable formulas for making glass. There is a formula for making a glass that, when immersed in the water of a lake or stream, immediately registers a measurable voltage; with that reading, the acidity of the water can easily be determined. Such glass electrodes have a wide range of applications in chemical procedures today.

And from another formula came a new nonreflective glass, one like that installed on street level at Forty-ninth Street and Rockefeller Plaza in New York City, so pedestrians could look in on NBC's *Today* show. Schott has it, a glass that reflects just an eighth of the light reflected by ordinary glass. Such glasses, also made by Corning and others, are snuffing out the glare on the television screen so that Peter Jennings need not appear with the reflection of a potted plant superimposed on his forehead. Some new glasses being made today contain greatly reduced amounts of iron, resulting in a clarity without the pale green tint of ordinary flat glass. An object's color isn't distorted when viewed through this glass, and, if anything, it lends a sparkle to what is seen. This is the glass of which I heard a woman say, "When you look at a pea through it, you see an emerald."

Glass as an agent for disposing of millions of gallons of high-level nuclear waste? That is being done. Glass to harness the power of photons and resolve images down to one two-millionth of an inch? Glass to mend broken bones? A dentist milling a glass-ceramic to make a set of dentures in the office? Yes to all of that. Glass has stamped the world of science with the vital necessity of itself.

"It has so many unique properties—its ease in shaping, its transparency, durability, and low cost—that I think it will be indispensable in the growing communications, information, and electronics industries," said William R. Prindle, a former vice president of the Technology Group at Corning. "For one thing, those industries require display systems for their computers and televisions, and glass—either as a picture tube or as a thin and flat sheet for a liquid crystal display—is the material of choice."

So the probing and the effort to swell the material's versatility continue, always with the pleasant awareness that the main ingredient, sand, is not likely to be exhausted since it comprises something like 14 percent of the earth's crust. (The specter is always there: the earth being cast into glass as the world ends by fire.)

Although the great body of technical advances in glass is from the twentieth century, interest in glass as science started nearly four hundred years ago with the discovery of the telescope by Galileo. Before that, there was little, if any, interest in glass as optics. But when this great instrument came into use, scientists gave strong attention to improving the optical properties of glass. Similarly, development of the integrated circuit, responsible for crowding our lives with gadgets that the noble Galileo would never have thought possible, has served in large measure to step up development of glass for use in these times of the microchip.

With television and the computer, calculators and video games, even with the screen at the supermarket checkout counter that flashes $13.95 for something chickenlike (anorectic and frozen)—with those and others, display, meaning the transfer of information to the user, is of primary importance. And that is being done with the use of glass and gases and liquid crystals. The technology used is gilt-edged, and it incorporates a tight embrace of glass with science.

Images are everything, for these are times of packaged

visual excitement. Push a few keys and the computer responds with a display on the screen, as a television and calculator also do. In the new information age, the emphasis is not so much on hearing as it is on seeing. We have come to expect "visuals," as photographers call them (unfortunately), to pop up all around us. It is no longer sufficient to hear the radio announcer say or read the newspaper report that it's going to be 74° and sunny today; we want charts and flow lines and a meteorologist who makes us laugh. Don Imus and Howard Stern, the biggest draws on radio, now have their shows broadcast on TV. With the flashing action of electronic games all around, when is the last time a child had fun squeezing two pieces of wood to have a monkey loop-the-loop on the crossbar? There is nothing wrong with any of that, of course, although the costs are high. And with each new display technology, more sophisticated than the last, they go even higher. The glasses being used in the systems are expensive because, for a liquid crystal display application, they must be of superior quality, free of impurities. The glass that has set the standard for that is the one made by Corning, identification number 1737.

Liquid crystals are organic polymers with long molecules. Such crystals are found in the scum of soap or a squeeze of satiny shampoo. Being a liquid, the compound should have the random molecular structure of a liquid, the same as glass has but shouldn't. A crystalline order is improbably present, however.

Placed between two pieces of flat glass, the rodlike molecules are free to move independently of one another, although they do tend to remain aligned, like geese in flight. It is with the introduction of electricity that the molecules become fully oriented, allowing the passage of light. The principle in use is not unlike that of the privacy glass made to turn either opaque or clear with the use of electrical current. But for displays, it is much more complicated in that the application of voltage must be selective. Each row of crystals is given an address pulse, and then those in a position to form the desired display are given voltage. Problems can arise when cells of uncharged crystals pick up some of what is called "cross-talk," or stray voltage, and that affects the contrast and resolution of the display. The color in the display is obtained by the use of phosphors.

The basic technology for flat-panel displays goes back to the 1960s, with development of the plasma display system. In that, gases incorporated into the system respond to electrical current by ionizing and emitting light, which is controlled for display use. Of course, the main tool for display is still the cathode-ray tube, with its electronic gun tattooing displays on the luminous screens of thick and heavy televisions, radar equipment, many computers, and other electronic devices. Technological development of displays over the past twenty-five years has concentrated on lighter, smaller systems capable of transmitting full-color displays of high resolution, sharp contrasts, and unsullied brightness to match or even surpass those of the cathode-ray tube—and do that using diminished charges of electricity.

Refinements to the technology have been made through the years until now there has come into being the active-matrix liquid crystal display. "The flat display business is a twelve-billion-dollar one today," said Dr. Steven W. Depp, who does his display technology research at IBM's Thomas J. Watson Research Center in Yorktown Heights, New York. "And most of the attention is now on the active-matrix liquid crystal display. It already offers a better image than the cathode-ray tube, if looked at straight on, but it is costly."

In the making of the active-matrix system, liquid crystals are once again sandwiched between glass. Only this time, the movement of current is controlled by a million or more transistors, applied in the form of a film. Each pixel, or element of the image, has its own transistor to control the current, and in that way, the cross-talk problem is avoided. Of course such transistors can hardly handle much electrical power, but, as Dr. Depp said, "It doesn't take much power to control a liquid crystal."

This is not so simple a technology that you can touch a soap bubble with a bit of wattage and expect it to flash the time of day. At the same time, it is technology that depends as heavily on control as it does on innovation. For one thing, the system is not going to work if the glass used does not meet certain specifications. It must be free of alkali. Nothing salty must touch the glass—not even salt so delicate as that of a tear from a baby's eye—for that will contaminate the thin-film transistors in the active-matrix display system. The glass must be light and thin, with an unblemished surface and of unvariable flatness. It is probably the purest glass being made today.

Such glass is made by what is known as the fusing process. This is how it works: Molten glass is allowed to flow into a trough until it overflows and spills down both sides, becoming stiffer as it falls. The two spills come together—fuse—at the bottom of the trough. Not a single human hand touches the glass until it has cooled and soldified and is ready for cutting into sheets about three feet square. The two sides that touched the trough when spilling over are buried in the interior of the fused glass. Not only is the glass flawless, very thin and precise, but it is also capable of withstanding high temperatures without expanding or undergoing other changes.

"Most of our patents on the fused glass process have run out," a Corning spokesman said. "People around the world are trying to do this, but it's not easy. It takes a lot of scientific art and a lot of patience. It's a novel approach to making glass."

It has taken nearly forty years for thin-film transistor display systems to reach the threshold of commercial application, and the reason for that is due in good part to the promise of financial return, or lack of it. Television sets with screens of nineteen inches and more continue to have elephantine rear ends because the cathode-ray tube, in addition to being excellent technology, is—and I use the term with a frisson of shame—"cost-effective." It wasn't until 1984 that transistors were used for a television display, on a screen of just two inches. The resolution was excellent.

RCA invented the technology, and Westinghouse and Seiko-Epson picked it up long enough to demonstrate that it can work, that it is indeed possible to put liquid crystals between two sheets of glass, then add electrical current to change the structure of the crystals and therefore control transmission of light. It is simply, and beautifully, an electronic venetian blind.

Still, that technology, and the technology for all flat display systems, sat on the shelf. It remained for the Japanese to pick it all up and go from there, visualizing, no doubt, that (because of form and power factors) flat displays are certain to be the dominant systems in the field of electronics before too many years of the twenty-first century have passed. Already, the NEC Corporation in Tokyo has introduced PlasmaX, a forty-two-inch television set with a flat display screen. The set itself is no thicker than a bookshelf. The system used involves light-emitting plasma gas between two sheets of high-quality glass. But plasma displays require a considerable amount of power, and so liquid crystals cannot be far behind.

And after that? There is on the horizon, far away, an emerging display technology that incorporates the use of more than four hundred thousand aluminum mirrors suspended over the memory elements of silicon chips . . . but that has a long way to go, since it hasn't even reached the stage where a decision is made to replace the aluminum with glass.

"We have the vacuum in the glass enclosure of the flat display screen, and the dream has always been, because of the volume of this vacuum, to squeeze it down to the size of a picture hanging on the wall," Dr. Francis Fehlner, who has conducted research into glass for Corning, said. "We'll have a two-dimensional display that looks like a picture on the wall. That's what we are aiming for with flat display." IBM's Dr. Depp sees it moving far beyond that. The same flat display that hangs flat on the wall as a television screen can serve as an interface with a computer. It can be a photograph album or a book for reading. Connected to a digital library of art, it can lend a Gainsborough or Monet to the room. It is in line with the movement toward an electronics superunit incorporating all the entertainment and communications devices in the home. It will be, no doubt, very complex, a lightning storm of digital flashes. And, by all odds, it will do absolutely nothing to simplify the use of it, such as playing just one selected track on a compact disc.

As natural obsidian, glass gave hunters and gatherers of food and fur material for tools more than a million years ago, and now, as flat display substrates of fine precision, it is about to do the same for those on a quest for the electronic future. Among materials in service today, only glass could have moved so effortlessly from cave to media room.

There comes word every so often of a new scientific use for glass, something esoteric that, even though it holds much promise, fails to make it out of the laboratory. It can be a patent-pending fruition of an academic's long work, as was the case of the glass beads that clean up oil spills. The idea belonged to Dr. Adam Heller, a chemist at the University of Texas at Austin, and it gained national attention in 1992 when Dr. Heller, speaking at a meeting of the American Chemical Society, said that by using the beads, it would be possible to clean up an oil spill the size of one made by the *Exxon Valdez*—eleven million gallons—in just three days and for a fraction of the cost that the Exxon Corporation had to bear in fines and cleanup expenses.

The hollow beads would be tiny, like grains of sand, and they would be coated with a material that is part titanium and part palladium. When the beads were dumped in the water, the oil would be drawn to them, forming clumps. With the glass beads acting as semiconductors because of the metallic coating on them, they would take enough energy from the sun to cause combustion and the burning away of the toxic hydrocarbons in the oil. Bacteria in the seawater would eventually dispose of what would be left of the oil spill.

There was no reason to believe the process would not work. It was all good chemistry. The beads would cost $3,000 a ton—a bargain. Five years later, Dr. Heller was still at the University of Texas, saying he was a very busy person and had little time to talk about his glass beads project. Was it stalled because of financial considerations? "I will send you some literature," Dr. Heller said (glass scientists usually maintain warehouses of article reprints, and so restrictions should be placed on such offers). "But no, it wasn't financial, not directly. The oil companies—they didn't want the use of beads because the bad color of the spill wouldn't go away immediately. They wanted instant visual gratification."

Dr. Heller still holds out hope for the application. Meanwhile, he said, he's deep into work to transfer the same basic technology to making self-cleaning glass. Again, the glass would be coated with a film of titanium oxide that will react in sunlight, causing buildups of dirt and other spoilers of clarity to disintegrate and fall way.

Even large organizations such as Corning and Schott have products on the shelf, scientific innovations in search of a market. There is sol-gel, for example, a process for making glass without melting. Certain organic compounds are mixed with water, and that results in a bonding of different atoms and the formation of a gel that becomes a glass when heated. It is a ten-year-old invention without an application.

Dr. Gerald S. Meiling, director of science and technology strategy at Corning, dropped a glass tumbler on the floor of his office. It didn't break. Made of three layers of glass, a low-thermal-expansion type sandwiched between two layers of high-expansion-type, the material has extraordinary strength for use in the kitchen. "I took twelve of them home ten years ago, and we still have ten of them left," Meiling said. This glass is not something clunky and scratched, like that found with the table service at eateries where the dinner music is the sizzle of grease. It is suitable for use when company calls, and by all appearances, it should have been a successful offering on the glass market.

But the application of science to glass can be expensive. To take three pieces of glass, one of which is of a different composition, and join them together so that the three-layer stack of stresses acts like a tensing of all the body's muscles into a shield of toughness—that requires a great deal of work. The glass was priced to reflect all the work that went into the making of it, and the consumer said, "no, thank you." The question can be asked: Why do all the work in the first place?

Glass holds almost Circean appeal for scientists, and so they binge on the material for research even if the cost of doing so will not justify the results in terms of customer acceptance. There are so many things they can do with glass that it has become for them like chicken stock for a chef. But the call for financial accountability is being heard, and while there will continue to be misses as well as hits in the laboratory, they will not drag out so long.

There have always been efforts made to do something wondrous with glass, to takes the transparent, sand-based material and make of it a flash card of communication for the scientist and a medium aswirl in the dance of light for the artist. And there is something for writers as well, a metaphorical device good for giving body to potentially spongy passages: "Yet what composed the present moment? If you are young, the future lies upon the present, like a piece of glass, making it tremble and quiver. If you are old, the past lies upon the present, like a thick glass, making it waver, distorting it." So wrote Virginia Woolf in the essay "The Moment: Summer's Night."

Do not dismiss the possibility that someone even now is reinventing glass—perhaps a member of the Murung tribe in the Chittagong Hill Tracts of Bangladesh, sitting on the shore of a lake there, deep in thick growths of teak and banana trees, putting fire to sand and soda and lime, and thinking no one else has ever done this before. If not there, then maybe in the rain forest of Mato Grosso in Brazil, far, far off the road, back where Aroeira Indians, who are wise in the ways of the tapir, carry spears tipped with a poison that is a highly effective anticoagulant; it could be happening there, the rediscovery of glass by an Aroeira who doesn't have any idea that he or she has come thirty-five hundred years late to this.

It was as recently as the late 1970s, in the Afghanistan city of Herat, that glass was being made as if it had just been discovered. Two men working in a small room had for their equipment a furnace made of mud bricks and a collection of tools no different from those used in ancient times. They used wood for fuel, and somehow they got the fire hot enough for the melt. For the silica, they had only pebbles from a riverbed. It all worked. They made glass.

Science has taken glass out of the mud-brick furnace and put it in a computer-driven one with a platinum crucible. Though often held in mind as a material with plebeian roles—windowpanes, light bulbs, glasses to drink from—glass has always been more than that, but it took science and post–World War II technology to pluck the jewels of versatility from the rest. Glass as fiber optics, strands of glass the world will wear like a necklace to change the ways of communication—that's one of the best.

Chapter Six

Good-bye Copper, Hello Glass

Think of this, of glass encasing light, and the light a carrier of silent voices and much of what we know and want to know, and all of it together a tool for adjusting the world to the fast time of science in the twenty-first century.

This glass is called fiber optics, and with the ever-rising need in the world for better ways to move communications and information, it has been brought forth as the great provider. Already it has heavily eroded the omnipresence of copper wire, and the day draws near when the last of the familiar color-coded wire, bunched to the thickness of a ship's mooring line, is disconnected from its terminal and pulled from the ground. Then the glass fibers will have reached the home, bringing voices to the telephone and quite likely also pictures to the television and libraries of information to the computer.

To see fiber optics at work and come to know their wonders, go to a rural area, out where lost roses grow in graveyards, and visit the school. There is a better than slight chance that the students there will be sharing classes—a mutual struggling with algebra and past participles—with students of other rural schools in the area. They will be doing that through the use of television, an educational video hookup through which images and audio are carried by fiber optics. The schools may be twenty or even fifty miles apart, but they are joined together by glass buried under fields of wheat and corn, and rivers that rope through old growths of cottonwoods.

As of now, glass fiber optics are rapidly replacing copper wiring for use as telephone lines. Atlanta-based BellSouth, for example, now has most of the lines in its nine-state system switched over to fiber optics. However, the new carriers, while in use as trunk and feeder lines, have not yet been routed into individual homes. "Fiber optics are in place to where they can be regarded as the interstate highway of communications, but copper wiring remains, for now, as the neighborhood street," a spokesman for BellSouth said.

There are cost factors at work. Although the expense and increasing shortage of copper gave some urgency to the introduction of glass fiber optics for telephone use, the cost of converting a system can be staggering (by the time all local telephone companies in the United States make the conversions over the first quarter of the new century, the cost could well run to $100 billion or more). The telephone companies—no surprise here—want to transfer most of that burden to their customer. Since the rates charged by telephone companies must be approved by government agencies, the road to recovery of costs for replacing copper wiring has not been all smooth.

Aside from use in communications, glass fiber optics now have a ubiquitous presence in modern technology. If set down in a straight line, the amount of the material in the ground would reach for about thirty million miles; adapted for dental use (to move away from analogies that figure in the distance to the moon), that would be enough floss to last everyone in Norway for a little over two thousand years. And there is much more to come, for the device is a scanner and a monitor, a probe and a guide. And yet it is a seemingly uncomplicated filament, thin as a thread, but superstrong and flexible enough to be wound around a finger. It is glass once again being deceptively simple in its role as a mainstay for high-test science and technology.

Fiber optics are replacing metal wires for a number of reasons in addition to the cost and growing shortage of copper, but more than anything else, use of glass offers a far greater capacity for handling information—more than twenty-five thousand times as much—and can transmit it at a speed that leaves copper wire at the post. And there is the important advantage of being lighter—indeed, while a mile of fiber weighs only between four and five ounces, it would require more than thirty tons of copper wiring to match the transmission capacity of that feathery amount. Take a single hair from your head and hold it up and know that light is being driven through strands of glass as thin as it, and that those flashing pulses are the heartbeat of an emerging communications giant. These are the same light pulses in the same glass fibers that will one day replace electrons to store programs and process information in optical computers.

Today fiber optics are being used to obtain a clearer image of smaller and smaller objects than ever before—even viruses. A new generation of optical instruments that can provide detailed imaging of the inner workings of cells is appearing. Called near-field scanning optical microscopes, they can harness what one scientist calls "the power of photons" to resolve images down to approximately one two-millionth of an inch. They see not only what's small, but also what's in the dark. Millions of fiber optics can be bundled together to amplify light a hundred million times. Such components are used to turn night into day through their use in special goggles. As part of a rifle sight, they allow a shooter to see his target even on a moonless night.

The surge in the use of fiber optics has given new economic strength to the glass industry in the United States. It all began in 1966 when scientists reasoned that it should be possible to send a beam of light from a laser through a glass fiber. Getting the transmission started would not be a problem, but retaining the light as it moved farther along would be difficult. If just 1 percent of the light could survive a journey of one kilometer, or a little over a half mile, that would be counted as a success. In 1970, Corning announced that it had done just that. "That spawned the present age of telecommunications," said Dr. Donald Keck, one of three Corning scientists responsible for the light-retention breakthrough.

It was known long before fiber optics were developed that light could carry communications. Certainly, Alexander Graham Bell knew it as far back as 1880, when he introduced his photophone. For this technology, the inventor of the telephone borrowed on the mischief of a youngster who tries to reflect sunlight with a handheld mirror in a way that it strikes the eyes of someone else. Bell used vibrating mirrors in an attempt to move sound with sunlight. It worked somewhat, but it fell by the wayside in the shadow of the telephone.

Not until eighty years later did the world gain the laser, which was to become an instrument of broad use, not just a threatening weapon in the hands of the evil Auric Goldfinger. It would provide the rapidly pulsing light—billions of pulses a second—necessary to set off a fiber-optic revolution in communications. The word *laser* is an acronym for "light amplification by simulated emission of radiation." The laser acts to excite atoms which in turn produce light that is amplified. The light is made directional by the shape of the laser device, allowing it to pour directly into the fiber optic where it then travels through the impurity-free glass without being scattered. The light rolls in a wave that is made to modulate, and on those pulses the communications are carried.

The laser starts to pulse light when coded electrical signals are introduced into the system. In effect, the electricity is converted into light at the head of the fibers, and on the other end, the light reverts back to electricity to power the communications receivers. The light—pulsing, say, two billion times a second—cannot be seen because it is in the infrared portion of the light spectrum, acting like a semaphore rather than a steady beam. Each pulse is carrying a bit of information, such as a part of a telephone conversation.

One of the most astonishing achievements along the road to a successful fiber-optic communications system has been the essential exact alignment between the laser and the tip of the fibers. The light of the laser must hit its target straight on, zero in on a bull's-eye of a size calculable only in millionths of an inch. The light cannot be "mushy," such as the light from a bedside table lamp; it needs to be of razor-sharp focus, or what is known as "coherent" light. The light zaps its cargo of information through a target in the tip of the fiber's core that is more than four hundred millionths of an inch wide. It is all a masterwork of precision.

Since the first successful fiber optics, when 1 percent of the introduced light survived for a distance of one kilometer, refinements have been made; now, 80 percent of the light will survive transmission of one kilometer (passing through ordinary eyeglass lenses, that 80 percent of light would move for no more than three feet). To travel long distances, there must be amplifiers along the way to give lift to the light and prevent it from reverting back to electricity before reaching the end of the fibers. It is essential that the light not be scattered and therefore suffer loss. That will happen if it runs into an impurity in the glass, just as a skier on seemingly unblemished snow might suddenly be bounced by an unexpected mogul.

The success of the system has depended on the glass from the start, glass not made by the conventional melting process. This glass is made from ultrapure vapors of silica and other ingredients of glass. Treated with fire, the vapors form soot, which is deposited on a rod. With the rod then removed, the soot preform is placed in a furnace where, with the water vapor removed and the hole left by the rod now gone, it is sintered into solid glass, which is then placed in a draw furnace. Melting, the glass is drawn into fiber optics— continuous strands of the royal jewels of glass, purer than any other and stronger than all but the reinforced. A cladding, also of glass, is added to the fiber, in addition to a protective coating of acrylate. With all of its armor, the glass has achieved a theoretical tensile strength some six thousand times stronger than that of copper. A three-foot-long strand of finished glass fiber would have to be stretched a little more than two inches before it would break.

The goal is to have it take from ten to thirty years for a crack in the glass fiber optics to grow to any proportion where it will cause a problem. Working in the laboratory, scientists have drilled holes in specimens of the glass and then set cracks to start by applying stress. "We are in effect growing cracks, and we want to be able to have them grow very, very slowly," a research technician said. Again, as with all glass, breakage is the bugaboo, but with fiber optics, it takes much longer.

The fiber optics of a communications system function in a world of radical reduction, far, far down to where measurements are in nanometers, down to where the dot over an *i* would be of a size to reckon with. To splice fiber optics, the glass is melted and then fused, and so it will come to be that the repairman will have to work more like a surgeon than a splice-and-tape tender of copper wire. No longer will there be men or women of the phone company like the one I once came across in my neighborhood.

He was down in one of those holes in the ground where they go to make repairs after erecting a portable fence around the top of the hole to keep people and cats and dogs from falling in. He was singing, and his voice carried up and out and down the street for a way: "Now the little blue wire goes here, and the little red wire goes there, oh that's where they go, where they go, where they go. Blue wire here, red wire there, that's where they go." He was a person who descended into those holes like a happy mole, and his likes may seldom be seen again once the wire, with its high maintenance needs, is all gone.

But there is little cause for worry that neighborhoods will suffer loss of character because of that. Utilities always find a reason to dig holes in which to deploy workers.

Once in place, a fiber-optic communications system, with its massive carrying capacity, will almost certainly bring about a deluge of new options for use with the computer, television, telephone, and just about any device that buzzes or lights up when turned on. Douglas Hall, a fiber-optic scientist with Corning, has looked ahead: "Right now we use fiber optics to transport data. I can visualize the day when fiber optics play a role in the manipulation of that data."

His reference is to the optical computer, and the storage and processing of information by means of laser-triggered pulses of light instead of electrons. As with the telephone line, the light will travel through glass fiber optics, but instead of re-verting back to electricity at the end of the fibers, the communications-carrying signal will continue on as light and be capable of storing and processing the information in a computer. The optical computer will function with such dazzling speed and efficiency as to make faint competition of the fastest computer now available.

The idea of an optical computer is not a new one, but only in recent years has the development emerged from the conceptual stage. "A lot of new devices for optical computing are on the street now," said Donald Chiarulli of the University of Pittsburgh, who has done extensive research in the field. "But it would not be a good idea to build a computer that's a hundred percent optical. There will still be room for electronics—for presentation of the information."

In January 1993, two researchers at the University of Colorado at Boulder stood in attendance at a mind-thumping array of wires and switches and snarls of circuitry; it was a compound for photons, the world's first multiservice optical computer. Harry Jordan, a codeveloper of the machine along with Vincent Heuring, explained that data is recorded and processed in the computer by means of infrared light pulses from a laser circulating through glass fiber optics, and then being routed through switches made of lithium niobate. The fiber—a two-and-a-half-mile-long strand—is tightly wound around a spool, and each information-carrying light pulse travels through the full length of the fiber at a speed of 186,000 miles a second.

Five years later, the computer remains a prototype, and both Heuring and Jordan have moved on to other things. They feel they did all that they set out to do. "We really didn't develop the computer for commercial purposes," Heuring said, "but rather as a way of getting information on computing at the speed of light. We developed a whole new body of theories on that—how you treat the light pulses and so forth." They succeeded in taking optics into the computer to serve as the force for storing and manipulating the data. There was one earlier optical device for computing, but it had to receive instructions and information from an electronic computer.

"We proved it would work," Jordan said. "There will always be a need for electronics in the computer, but more and more optics are getting into the machine." Neither he nor Heuring visualizes the optical computer becoming a standard desktop machine anytime soon. "Maybe in a hundred years," Heuring said. "The first application will be for telecommunications."

What can be said of such speed in a device created by humans? The same that was said of reduction when the Lord's Prayer was first inscribed on the head of a pin? There is a certain triteness to citing the speedy movement of information today, as the life span for computer superlatives is tenuous. But when the age of photons is upon us in full, there will be cause for awe. It will be a time for dramatic achievements for both computers and voice communication. Of course, it can be said that there is no need for the transmission of more than ten million telephone conversations in just one second, all routed through a single glass fiber. It does seem to be overkill. How often—think about it—is a call delayed because all circuits are busy? Sometimes on Christmas and Mother's Day, yes, but otherwise the beckoning dial tone is there to fill the ear like birdsong in spring.

It remains to be determined for certain how glass fiber optics, when buried in the ground, even in the embrace of cladding and plastic coating, will fare over the long term when the environment is one laced with fertilizers and other chemicals. Heat and humidity are other potential enemies of the fiber, as well as, no doubt, certain wildlife creatures. In the south of the United States, thought must be given to the fire ant, a voracious insect packing deadly venom and a deviant liking—more like a lust—for electricity. An invader from

South America, the fire ant arrived in Alabama in the 1940s and has by now worked its way across most of the South and a good portion of the Southwest, always on the lookout for a good electrical field. They will chew through insulation to get to voltage, and although some are dispatched in the process, most happily stuff themselves on shock. The fire ant has been a source of deep trouble for utility companies, although they hesitate to admit that.

Fire ants, of course, may not like glass at all. The idea of eating the material has always been gist for tales of horror: the avenging wife who sprinkles crushed glass in her abusive husband's oatmeal, the jagged shard concealed in a Halloween apple. At the same time, there is Michel Lotto of Grenoble, in France. Although it remains to be determined if he has a taste for fiber optics, he has eaten his way into the Guinness Book of World Records with meals of, among other things, crystal chandeliers—six of them over the years.

Of all the showcasing of fiber optics as a device for reconnoitering the future, not many and maybe none have been as spectacular as Georgia's new Transportation Management Center, a $140 million system to combat traffic congestion in the Atlanta area. In this, glass has been put to use to help a major city save itself from an onslaught of motor cars and their attendant pollution. With two major interstate highways—I–75 and I–85—merging to push as one through the heart of Atlanta, the need for relief is acute. A motorist can go mad on that road, and on the one that circles the city—I–285, a loop more threatening than the hangman's noose. In 1996, with opening day of the Olympic Games approaching and the certainty of even tighter gridlock, it seemed that the end was near—a day when, perhaps, a skinny guy who trained by chasing lions across the African veldt wins the five-thousand-meter run while the host city ceases to function, choked to death by one too many Grand

Marquis on its seasonal pilgrimage between Florida and Michigan's Upper Peninsula.

The system was put in place at a time to coincide with the start of the summer Olympics not only because of traffic, but for the visibility advantage as well. Nevertheless, it is a permanent fixture, "a lifetime thing that will grow and improve through the years," as Kathleen Barron of the Georgia Department of Transportation said. Not only that, but the federal government, whence came 80 percent of the financing, is looking at this as a foreshadowing of a smart urban transportation infrastructure of nationwide scale.

They went into the heart of the urban monster that Atlanta has become and laced the ground with 188 miles of glass fiber, laying it out at a depth of from thirty-six to forty-eight inches along the interstate highway where it crosses the city, and also along arterial roads. They installed close to four hundred cameras for surveillance of the autoscape, some of them positioned on poles as high as one hundred feet. Over $4 million worth of electronic traveler-information kiosks were located along the highways as well as on city streets. Changeable message signs were raised above the traffic, there to offer advisories with marquee appeal. To regulate the merging of vehicles onto freeways, ramp meters were installed to allow access to the road at a measured rate; that lessens the danger of accidents and sharply reduces the clogging of cars and trucks that often occurs at entry ramps.

And, using fiber optics, they linked everything up over a five-county area—stoplights and roadside advisory signals, emergency service alerts and more. Everything comes together at a computerized control center, to where the cameras send their visual readings and where traffic flows are adjusted through instantaneous timing adjustments of most of the city's 840 traffic lights. A button is pushed, and a command goes out for a highway message sign to advise motorists of what's ahead, and, if it's bad, how to avoid it.

There is a garnish of esoterica on the system, such as a hotel guest's being able to call up customized traffic advisories on the in-room television set, and computers small enough to fit in the hand and still offer up traffic and other city-related information. Much of that has fallen away since the close of the Olympics. "It's not that they weren't successful," John Shirley, technical and communications manager for the Transportation Management Center, said. "The handheld computer, for example, was very successful, but it was very expensive to operate. It came down to being a matter of economics."

State and city transportation officials stress that the system is not meant to be a cure for Atlanta's traffic problems. "We hope to manage the problem this way," Kathleen Barron said. "To cure it will require lifestyle changes, meaning more people in fewer cars." Atlanta is not likely to ever again be crinolined and gracious, and the same can be said for its chances of achieving traffic abatement. So the city is doing what it can: balancing its hope on the hair-thin strands of glass that lie buried in the red Georgia clay. The federal government would like at least seventy-five other large cities to do the same.

Members of state legislatures and city councils have paid close heed to proposals for linkups of schools and other public facilities by fiber optics because the federal government is willing to bear a good-size portion of the cost. And so the state of Iowa, with $10 million from the Department of Defense in hand, has tied its National Guard armories together; now, the voice of central command rides the light pulses to wherever there's an armory, if not to alert the troops, then to confirm the facility's availability for Tuesday-night wrestling.

In addition, Iowa's fiber-optic network brings together hospitals and schools and universities, as well as some other public facilities. The network became active in 1983, and since that time, at least five other states have embraced fiber optics as a means of improving public services. The thinking is that if, for example, the level of education rises as a result of linking schools, there will be a workforce with improved qualifications, and that will attract more industry to the state. Whatever the motive, students and patients can only benefit from Iowa's pioneering showcase of information routing.

For students of rural schools on the hookup, the time has passed when instruction was more or less confined to what the textbooks offered. Now, the strands of glass buried in the ground are bringing excitement and mental stimulation into those rooms of learning: A guest lecture by a university scholar two hundred miles away energizes the sixteen seniors of a small school with new insights into the genius of Shakespeare. For work on a term paper, the fiber optics trace a road to the library stacks at the state universities, where a vast store of reference material is available for call-up.

Boosters of the network in Iowa like to cite the time not many years ago when eighth-graders in Mount Ayr, in the southern part of the state, wanted to hear what James Van Allen had to say about space flight. (He discovered the belt of radiation that girds the earth and now bears his name, the Van Allen belt.) He was eighty then and not anxious to make the long trip to the school, so he went instead to a facility at the University of Iowa, where he is a professor emeritus, and there enjoyed a video and voice exchange with the students via fiber optics.

Other states are following Iowa's lead, and as more and more fiber is laid in the ground, in more and more states, the web spreads and spreads to where one day it may become a national hookup, something resembling the vaunted superhighway of communications as proposed by the Clinton administration in 1993. Since that time, uncertainty seems to have taken control of that project.

"The picture has changed enormously since 1993," said Mitch Kapor, a founder of Lotus Development Corp., a software company. "Back then, common thinking in government and within the telephone industry was that government would have to fund, in some fashion, the development of a national information infrastructure. But today, of course, the Internet is where the action is." For the Internet to serve as the new data highway, there would have to be considerable upgrades, including availability of adequate "bandwidth," or carrying capacity. If existing telephone lines are to be used, they will have to transmit data at much higher speeds and with greater capacity. With both cable and telephone companies in the picture, there has been controversy, and, in general, no warm and fuzzy blanket of cooperation has swaddled the project.

"At the start, most telephone companies thought television was going to be the driver for the proposed data highway," Booker Tyrone, of the Southwestern Bell Technology Co., said from his office in Austin, Texas, a city as high-tech trendy as any. "It would be video on demand. It turns out that television has now been surpassed, and that the Internet is the driver. It takes a lot of bandwidth to transmit video, and so it needs to be compressed; it looks like fiber optics can do that. It can also handle data. Fiber optics is certainly the technology of the future."

If, in the end, it must be done completely with newly laid fiber optics, the data highway will cost, by most estimates, at least $400 billion. And that is one reason why it is likely to evolve (if, indeed, it goes forward at all) as a hybrid, incorporating some telephone lines and television cables, and a lot of fiber optics—enough glass laid in ditches, you would think, to set the earth to tinkling.

Glass fibers aren't just a device for carrying a voice, a picture, or, say, the three thousand or so pages of *Webster's Unabridged Dictionary;* for the military, they are an important component of sophisticated weaponry. Because they can see places where gunners cannot, fiber optics are being used to keep weapons on track to the target. Although still under development for the government by the Raytheon Co. of Lexington, Massachusetts, the Enhanced Fiber-Optic Guided Missile (EFOGM) looks to be like nothing less than a slice of wizardry from the cutting edge of modern weaponry.

The launcher is mounted on an army vehicle, and from there the missile rises, a slow mover with a tail of fire. It moves on and on, mile after mile, on the trail of a tank or a helicopter. And all through the flight, a spool wound with nine miles of glass fiber is letting out the filament until it trails like an umbilical cord between the gunner on the ground and the weapon. Once in the area of confrontation, infrared light in the fiber optics picks up images all around and sends them back to the gunner, flashing through the thread of glass as it rides on the wind. And with that, he or she can pinpoint the target and then position the mission on deadly track.

Although the costs are high, development of weapons using fiber-optic technology is expected to be pushed by the military. New weapons beget new defenses, of course, but it remains to be seen how this challenge will be met. (A chilling thought: What if some rinky-dinky state—Dagestan, maybe—achieves fiber-optic capability?)

And what of fiber optics being used to channel sunlight into a building in such a way as to provide all needed illumination with an efficiency five times higher than that of incandescent light? The technology is at hand, and not only will there be light, but heat and air-conditioning as well, once the light's infrared and ultraviolet components are harnessed. In reporting on this issue of fiber optics, as pioneered by a graduate student at the Massachusetts Institute of Technology, the journal *Technology Review* cautioned that "the only hitch is that the fibers could not be dimmed and there would be no way to turn off unneeded individual lights."

The waxy fire alarm does not have that problem. Here again, the glass fiber has been strung like a seamstress's thread through a cloth of ingenuity. Light pulses through glass fiber optics in a cable, lashed to a cylinder of wax. Should a fire break out in the general area where the alarm is located, the wax expands from the heat and puts pressure on the cable. The path of the light through the fibers is disrupted, and so the light scatters, causing a computer to set off the fire alarm. With the fire extinguished and the heat abated, the wax shrinks back to normal size, the light resumes its uninterrupted laser pulses, and the alarm shuts off. Goldbergian to a point but, overall, simple and reliable.

With its capability to function not only as a carrier of information, but also as an eye and feeler, fiber optics are rich beyond avarice in applications.

At about the same time thirty years ago that Corning was announcing it had succeeded in developing glass fiber through which 1 percent of introduced light could move for at least one kilometer, I was in Alaska to observe the onset of the great oil recovery activity on the vast North Slope. Oil wells were being drilled and capped, drilled and capped. And while the North Slope has changed, though still a brittle place of ice-heave sculpture—seventy-six thousand square miles in the grip of a silence sometimes broken by nothing louder than the fragile squeaks of lemmings—the technology for drilling for oil has more or less remained as it was. There are strong indications now, however, that the fiber optics are about to go down to where the rocks are soaked in crude, down in the sea to the dark realms of oil reserves.

It is not an easy task to find oil and then get it out of the ground. And finding ways to simplify and speed up the work has not been a robust success. As much as anything, new sensing devices have been needed—something to record temperatures deep in the earth without having to stop drilling while a measuring device is lowered down the well on a wire. It is dangerous as well as cumbersome, since an electrical spark can set off an explosion and fire. Finally, in this way the temperature can be taken only once during the procedure, and therefore must be repeated at intervals.

Tracking temperatures is important in an oil recovery operation since steam is injected into the ground to dislodge the oil from porous rock. The movement of the steam must be monitored to make certain the vapor reaches the rock, and the way to do that, of course, is by taking temperature readings. Fiber optics can achieve this without interruption of drilling, without danger of an explosion and fire, and do it on a permanent, single-installation basis. The fiber, clad, coated, and wrapped (and all of it together still not thicker than sewing thread), is sent down as far as, say, two thousand feet in a wash of water through a small tube of stainless steel.

"The glass of the fiber optics changes optical character with changes in temperature, and that in turn causes changes in the wavelengths of the light that's being pushed through by laser," said Robert Herning, general manager of strategic research for the Chevron Petroleum Technology Co. "All of that is recorded, and the information, being digital, can be sent directly to engineers, or around the world for that matter. It is still experimental, but there are tremendous applications for the use of glass fiber optics in recovering oil."

Chevron, a company with a liking for high technology, has been a leader in this research. The company's engineers have conducted field tests not only for temperature monitoring by fiber optics, but also for using the fiber as a sensor for pressure and as an aquatic sensor to determine if the flow into the deep well is indeed oil rather than water. "Fiber optics provide us with the ability to monitor changes in the reservoir. You can have water flowing in to replace some of the oil, and without a sensor you wouldn't know that. Again, the optic responds with changes in the light waves, only this time it's brought on by flows of water as opposed to those of oil and gas."

While pointing out that the technology would be costly, Heming looked at the long-term picture. "Compared with what we do in oil wells now," he said, "fiber optics could be rather cheap, considering the information they could provide us." In addition to its golden potential for use in offshore drilling, the fiber may finally give companies the ability to do what they've been wanting to do for many years: increase the average rate of recovery from an oil field above a measly 40 percent.

Fiber optics as a sensor are an instrument of dazzling versatility. A change in the light signal can flag the occurrence of a structural flaw in, say, a material of an aircraft. Even before the flaw develops, it can measure the amount of strain on the material. The fibers can go where neither man nor woman dares to venture—into spaces made radioactive, there to signal whether all is up to structural standard. Fiber optics measure as well, relaying data on vibrations and on slowing and acceleration.

And when the thunder of an avalanche shakes the floor of a bold canyon or snowy valley, fiber optics are brought in to assist in the search for victims buried in the pile-ups, as was the case in the summer of 1996 at Yosemite National

Park in California. At first just one rock fell, but that rock was close to four hundred feet wide. It plummeted two thousand feet, shattering into a rain of granite. Rescuers, using fiber optics to relay images from deep within the dust and debris, looked for survivors. It was bittersweet fortune that only one died.

The role of fiber optics in the saving of human life goes back many years, and yet it is still evolving, still supplying the medical world with capabilities for stunning new procedures for healing. If glass can give sight for a missile gunner's aim, it can also guide a surgeon through the follicles and around the ducts, down deep where hidden agents prey on life.

Chapter Seven

R$_x$ for Better Health

How many woeful eyes have looked down a runny nose to see a piece of glass sticking out of a mouth dry with flu? Sometimes it seemed you could hear the mercury encased in the glass practically boiling as it rose above the 100° mark, up to where your mother ("Keep it under your tongue") clapped a hand on your sweaty forehead to confirm the reading.

Glass is a material of wide use in medical practice and research. But as housing for a liquid in a clinical thermometer, it surpasses all other medical applications in age. It has served this purpose many years, at least back to the first half of the eighteenth century, when the first sealed thermometer was produced by Daniel Fahrenheit—all that shaking of the instrument, those half-turns of it to get a reading, the joy of a 98.6 in the morning after a night of 104.

The role of the glass thermometer in the development of modern medicine has not been restricted to certifying a fever. Rather, it stands as a symbol of the enlightenment that raised science from the dark and stagnant bogs of medical theories in the sixteenth and seventeenth centuries. It was the thermometer that stamped out such beliefs as that of a Scandinavian having a lower body temperature than someone living where balmy breezes waft among the coconut palms. It was the thermometer that provided a way by which the condition of the human body could be measured against an exterior standard, thereby severing reliance on the Galenic theory that all good health depended on a proper balance between what the body took in and expelled. From there, medicine began its turn toward functions of the body—breathing, blood pressure, temperature—for indications of illness.

Before Fahrenheit, both Galileo and a Venetian physician named Santorio devised instruments to measure temperatures, and even those early models, called thermoscopes, were usually made of glass. But it took the German physicist for whom the thermometric scale with a reference of 32° for freezing and 212° for boiling is named to seal mercury in glass and so provide an instrument still in use today.

Now digital instruments are available to take human temperatures, sometimes via the ear—a procedure that for some reason seems obscene, although clinical thermometers have been in worse places. The ageless glass thermometer—four inches long, and thinner than a chopstick—may soon be a rarity, but the use of glass for medical purposes seems destined not only to continue, but to swell with importance. No longer is the material's role in the field limited to containers and vessels for laboratory work. Glass in our time is itself becoming an agent for healing.

Envision a glass microsphere, a bead one third the thickness of a human hair, that has been put into a nuclear reactor to become radioactive. It is injected through a catheter into the artery that carries blood to the liver—in this case, a cancerous liver, a condition almost always fatal. The beads—five to ten million of them are injected at the same time—

follow the blood to the liver, and because a tumor attracts a heavy flow of blood, it is there that the radiotherapy glass settles. The beads are insoluble in body fluids, and although they remain in the body, no harmful effects of their lingering presence have yet been detected.

"In this way," said Dr. Delbert E. Day of the University of Missouri–Rolla, "the radiation can be concentrated in the area of the tumor and deliver seven to eight times the amount provided by routine methods." If applied externally, hit and miss, that amount of radiation could be fatal, and yet with bead application, the patient rarely has a side effect more serious than a slightly elevated temperature. "The beads are put into the artery and they are at the liver almost instantaneously—that's it, it's done," Dr. Day said. "The patients do not lose their hair, they don't feel nauseous. The beads lose all radioactivity in two to three weeks."

Dr. Day, a coinventor of the bead treatment, found that irradiation in situ, as opposed to external application, not only results in heavier dosage, but also causes less damage to healthy tissue in the region of the tumor. The glass of the bead contains rare-earth elements, such as yttrium, but it is free of alkali. No other material is so well suited for use in this procedure. The body is not hostile to the intrusion because, for one thing, the glass is free of impurities and not toxic. No significant amount of radiation escapes during the quick trip through the artery.

Although the hepatobiliary procedure has been approved for use in Canada, it is still under review by the U.S. Food and Drug Administration; it takes between fifteen and twenty years to gain FDA approval for a medical procedure involving an implant. Sanctioned or not, it is recorded in scientific literature that twenty-four persons with tumors of the liver were treated with the glass microspheres during a research project at the University of Michigan. Almost all of them received a minimal dose of radiation, since the purpose of the experiment was to gather data on the toxicity (or lack of it) of the glass. Nevertheless, the tumors in all but eight of the twenty-four patients shrank or stopped growing.

With increased attention being paid to the use of glass for beta irradiation (which, in terms of power to penetrate human tissue, is one step below the commonly used gamma radiation), experiments involving treatments of diseased kidneys are now under way, but rather than looking for ways to save the organ, the aim in this procedure is to kill it. The radioactive glass beads would function as they do with the liver, but the dose of radiation would be much larger. A cancerous kidney is often removed surgically, but there are cases where cancer cells could be spread to other parts of the body if that is done. In this situation it is best to kill the kidney where it sits, and the glass bead, functioning as a vehicle for intense radiation, can do that more quickly and cleanly than any other method now in use.

You have to think how wonderful it would be if a person could have a handful of glass beads injected into the bloodstream and have them moving from here to there, stopping off, like mechanics converging under the hood of a car, to fix things. As one example, glass could blunt the arthritic dagger points between nerves and bones and tissues. With rheumatoid arthritis, relief from the death-wish pain can be had with the removal or killing of certain inflamed linings in the joints. Surgery can accomplish this, but it is a painful procedure, and a patient can lose use of the afflicted joints temporarily. In Europe, physicians are allowed to kill the tissues by injecting radioactive isotopes into the joints. The FDA will not allow that to be done in the United States for fear that the radiation might burn beyond its target.

"The radioactive isotopes are short-lived," Dr. Day said. "And there is the danger of leaking radiation. There is an interest, then, in a material that will overcome both of those problems and still be effective in killing the diseased tissue. So we have been experimenting with glass. We have put handfuls of glass in the body parts of rabbits, and it's kind of like throwing a handful of sand into your car's transmission; it didn't seem to hurt the rabbits." Dr. Clement Sledge, former chairman of the Department of Orthopedic Surgery at Brigham and Women's Hospital in Boston, said treatment of arthritis with radioactive glass beads is a "technically challenging viable possibility." The glass, he said, would have to be biodegradable, and would have to be injected with a syringe.

All the world has searched for a cure for arthritis for so long that any report of a possible breakthrough is met with a mixture of skepticism and renewed hope, but always with intense interest. By reporting once on an elderly man's claim of having concocted a home remedy for arthritis, I threw him, in effect, into the mouth of a demanding and suffering dragon. The mail came from all over the world, bags and bags of it, and it continued to arrive for years to follow, almost all of it from the afflicted seeking to share his cure. This is what I wrote:

Brooks Robinson is a boatbuilder, but he has little time for that now. At the age of 74 he has made what he believes is an important discovery. "Brother, I'm not bragging, but I've got something here no one else has ever come up with," he said. And then he told me.

"I had arthritis for 35 years. I went through the war with it, and they couldn't do anything in the service. I came back home and went to building boats, and I got down with this arthritis so bad I couldn't walk. So I sat down to studying, and I came up with this mixture. Within five months I didn't have any more pain. Where

> *the arthritis went and when it's coming back, I don't know."*
>
> *Among other things his medicine contains a petroleum derivative, Epsom salts, menthol, and iodine.*

If glass fails with arthritis, no doubt there will be other physical ailments suitable for possible application. For the diabetic, there may be beads filled with the cells that produce insulin, glass beads that let the insulin out while preventing the antibodies from getting in. An enlarged prostate strikes me as being game for the bead, as does, for Dr. Day, cancerous tentacles of a brain tumor that reach out to where surgeons fear to cut. In another development involving glass as a weapon against cancer, researchers at Alfred University are working with glass that is magnetically susceptible. Charged with magnetic energy, the glass can transfer heat to a tumor, causing it to die.

Glass is so biocompatible with the human body that it is less than outrageous to think there may come a time when replacement body parts are made of the material. Germany's Schott Glaswerke is prepared even now to supply glass capillary modules that can be used, according to the company, not only for blood separation procedures, but also for the purification of biomolecules and the anchoring of cells on surfaces. These membranes are made of highly porous glasses on which high temperatures and organic solvents have no effect, and in addition to being ideal for medical separation purposes, they work equally well for separating waste from water, and for storage of certain hazardous materials. When treated to react to the presence of some chemical by changing color, the glass can become an effective instrument for making a medical diagnosis; it can also signal a dangerous buildup of chemicals, such as formaldehyde, in a room. In some cases, these devices can venture into realms of medical esoterica, out to where freeze-dried enzymes play a role in the procedures. As an agent for filtering, such porous glasses seem best suited for an artificial kidney, but look for them first as filters in automobile and other engines.

Of course, it may be a safe bet that glass will never have such applications, but then who would have thought that it could be used to restore hearing by replacing bones of the middle ear? Who would have thought that glass is now ready, nearly five thousand years after it was first crafted, to be used as a binding agent for broken bones and for replacement parts of hips and knees?

As life span increases, so do the number of surgeries to replace all or parts of joints worn by age not to the smoothness of old rock, but to raspy deformity. The chains have slipped the sprockets in these old moving parts, and there's little left but pain. But there are many elderly men and women now who are out on the tennis courts and the golf courses, swinging rackets and clubs, and rejoicing in the fluid, creak-free workability of their new or repaired moving parts. There are others, too, with broken bones pinned together. Most are packing metal in their bodies—titanium and steel implants used to hold things together until the mending is complete. The bond is secure, but the metal presents problems, and so glass, once again possessed of utility to overflowing, is being brought forward.

Binod Kumar, a senior researcher at the University of Dayton (Ohio) Research Institute, points out that the compositions of metal and bone are not compatible, and while metal holds things together, its load-bearing dominance prevents the bone from getting the stress it needs to grow and develop. Secondly, unlike a glass pin, which is resorbed by the body, the metal implant does not dissolve over time, and that can result in the need for additional surgery to remove parts of it.

Fred Knight, a retired U.S. Army brigadier general living in Sarasota, Florida, has stainless steel rods in both of his rebuilt knees. They are attached to bone with heavy screws. "I can feel the metal, and it hurts," he said. Asked if the feel of the implanted metal reacts to changes in temperature, he said, "I went to Colorado in winter and attended a football game when the temperature in the stadium was twenty-six degrees. Both of my knees—the whole legs, actually—went numb. It took a full hour for them to thaw out."

Kumar has been working since 1982 on a procedure to use glass fiber as the bonding agent for broken bones and replacement parts. "I think we have overcome the early problems that we had, and that the use of glass for this can be very important to medicine," he said. "We are now into experimental testing."

He began with a consideration of the main elements of human bone: oxygen, calcium and phosphorus. Why not make a mending device for broken bones that incorporates those elements for compatibility, that can be made as strong as metal, and is nontoxic and flexible enough to be shaped for impeccable fit? So Binod Kumar made such a device; he made a glass fiber, encased it in a biocompatible polymer, and set out to show that it was capable of tying bone together and holding it like that until it fused. This glass at its making is dosed with potions to ensure that it will dissolve in the body, but not before the fusing of the bone parts is complete.

In other forms, glass that engages in biological activity with the body is being used now for the packing of jawbones following tooth extractions, and for the replacement of bones in the middle ear in attempts to restore hearing. This is called Bioglass, and it is an extraordinary material that not only stands in supportive attendance at a healing site, but reaches in and makes good things happen.

Bioglass is an amorphous substance, like solidified Jell-O, and when exposed to hydrogen it turns into a silica substance. Placed next to a tissue or bone in the body, it reacts to the hydrogen in that, and, as a gel, bonds to the human part. When Bioglass is applied to a cavity in, say, the jawbone, an exchange of calcium and phosphate ions occurs between the material and body fluids, resulting in the formation of crystals of bone on the gel that the surface of the glass has become. Collagen fibers have moved onto the surface also, and it is among those that the new bone cells start to grow—and grow until the glass implant is a compatible part of it all, withdrawing its support as the new bone grows stronger, and, when it's time, dissolving and resorbing into the body.

In 1985, Bioglass, a registered name under the proprietorship of the USBiomaterials Corporation in Baltimore, was approved for use in dental procedures, for mending facial bones, and also as a material for artificial sound-conducting bones in the middle ear. With that, Bioglass became an early player in the rapidly developing market for body implant materials. As an implant, Bioglass has the advantage of bonding with soft tissue, and that allows it to move with the tissue and not against it. Implants in the ear that are made of other materials often rub against the eardrum, but that does not happen with Bioglass; the prosthesis for the ear is very small. While there have been some problems with it, the rate of success has been high.

Glass that interacts physiologically with the body does not behave the way common glass does. "It is polycrystalline in structure, somewhere between the glass of the tumbler you drink from, and the beach sand you sit on," said James L. Meyers, president of USBiomaterials. "It can be made as a powder or solid, and, combined with plastic, can be used as a coating or as a substrate. And it not only bonds to both bone and tissue, but it jump-starts the healing process. The sequence of the healing events remains the same; Bioglass just speeds it up. There's nothing else like it."

In Japan, it has been reported, orthopedic surgeons use large blocks of a kind of glass that is biocompatible with the body for the reconstruction of spines.

It must be regarded as an event of dizzying potential: human cells growing on glass—and not only growing, but being stimulated in their growth by the material. With such technology, what's to stop its application as a filler for all the anatomical depressions left over from a time when medical care was not so complete? It may be a way to fill what mastoidal cavities remain behind ears, for example. Possibly, glass as a healing agent could be a blessing for the tens of thousands of children in developing nations who have been deeply scarred by falling into open cook fires.

Of course, not all medical uses of glass need be so serious.

On top of all else, bioactive glass has now been found to be a material that assists in the restoration of dentin lost from tooth enamel. That is no small matter, since hypersensitivity in teeth is a widespread problem, one for which sufferers spend over $200 million each year worldwide on over-the-counter products offering relief. In the United States alone, there are an estimated forty million people with sensitive teeth, a condition caused by erosion of the mineral coating on the teeth and exposure of root surfaces. When that happens, tubules of the teeth, where the sensitivity lurks, are also exposed. Drinking liquids of extreme temperature can cause doubling-over pain.

"For effectiveness, one application of Bioglass is equal to forty applications of the most popular treatment for hypersensitive teeth," James Meyers said. Bioactive glass used in dental procedures, as with the glass used to bond or replace bone and tissue, offers a bonus with its focused purpose: In addition to dulling the sensitivity, it shows indications of building a permanent shield on the teeth to protect against abrasion and damage from the mouth acids that feed on the teeth.

There are, of course, some weighty problems with the use of glass in the body—dissolving too fast or too slowly, brittleness and flaking, the reluctance of physicians and surgeons to abandon metals and screws—but the material as a player on the field of medicine will not go away. Bone grafts, hip and shoulder operations, and knee reconstructions are waystations for the aged who, as a whole, are traveling into lives longer than any known before. And when they get the use of this biocompatible glass down to perfection, there should be few materials to compare with it as a medical aid. Working in consort with other materials, it could be even better. That the human body can find elixir in something so common as a sand-based material must be incomprehensible to some, but look at the body itself: an empyreal creation, yes, but still mostly meat loaf and mashed potatoes.

Glass is also claiming a rightful place on the instrument trays in hospital operating rooms and the offices of physicians and surgeons. Fiber-optic glass is being snaked through the body to act as a lantern and a mover of images, and to probe and measure, and doing all of that with minimal invasiveness. Hardly a meeting of surgeons passes these days without a paper on a new fiber-optic procedure being presented. Raoul Kopelman, a chemist at the University of Michigan, has fashioned a glass fiber so fine—a thousand of them put together would be no thicker than a single human hair—that it can get through the membrane of a cell without causing damage.

"It's not yet in use, but it does work," Kopelman said. He explained that the fiber is a sensor, capable of detecting levels of chemicals, among other things. The glass is drawn out until the tip is a whisper away from nothing. Working under a microscope, the tip is treated with coatings that react to the presence of chemicals such as potassium and calcium. Then, with the tip inside the cell, light is directed through the fiber, and the way the light shows on the tip—the color, the brightness—identifies the chemicals in the cell and reveals the activity taking place. The probe can track second-to-second chemical changes in the cell, a procedure not now possible (other than with the use of Kopelman's probe, a study of chemical changes usually entails taking tissue samples).

One immediate assumption: The fine-tipped probe will no doubt prove a valuable instrument for determining the condition of an embryo. Kopelman confirmed that in experimenting with live rat embryos, he found that the glass fiber did not interfere with the normal functions of the cells. The point of the probe is so small and sharp that it comes as a painless intrusion. Once approved for general use, it may even be taken up for simple procedures to be done at home, such as taking readings on blood-sugar levels.

Today in operating rooms, microsurgery is being performed with the aid of not only fiber optics, but the optical glasses of specialized microscopes. I once observed a brain surgery at venerable Massachusetts General Hospital, one of the world's premier medical facilities, and the operating room was freighted with glass—glass as instruments, as vials and other containers, as syringes, as fiber optics for looking into the nooks and crannies of the brain that lay exposed before us. But there was more to this presence of glass than just utility: Along with the stainless steel and tile, it brought a sense of indestructibility to the room, as a counterbalance for all the gauzy and rubbery things.

At Alfred University, Professor Arun Varshneya is working to make surgical procedures with fiber optics even more effective with the use of stronger carbon dioxide lasers, while still keeping the incision tiny. Laser-assisted surgery is rather widespread today, with much of it done in the doctor's office. The instrument is particularly useful in the practice of ophthalmology and dermatology, working, for example, to zap a mole or erase a tattoo. Carbon dioxide lasers are often used for these surgeries, with the light being delivered by mirrors and lenses and articulated armlike devices.

Dr. Varshneya would like to see that strong laser light used to clean the plaque away from the walls of arteries. But that is not like removing a mole. The surgeon would not be physically at the work site, and so there would be a need for a flexible fiber-optic device capable of moving the midinfrared light to the plaque without much loss along the way.

Silica-based glasses—even those as pure as the ones in fiber optics for communications—cannot do that.

"We need to be able to transport eighty percent of this light a distance of one meter," said Dr. Varshneya. "With what we have now, we can get only one percent to go that far." The answer, he believes, is fiber optics made of chalcogenide glass. Such glasses are made from compounds of the chalcogens—sulfur, selenium, and tellurium—used in conjunction with other elements such as arsenic, germanium, phosphorus, and silicon. Such glasses are free of oxygen, are semiconductors of electricity, and are very efficient when it comes to moving infrared light with minimal loss.

The technology for putting all of that together is not yet fully developed, but Dr. Varshneya continues to work to find ways to overcome problems with the melt process for this glass and the drawing of the fiber. "We have already produced in our labs specimens of glass that are close to meeting all requirements," he said. "We are at the point where just a small amount of additional research effort could have a large payoff. This laser will allow physicians to perform more delicate surgery, such as literally scrubbing the plaque off an arterial wall. Conventional laser power may be able to ablate the plaque, but it may also perforate the arterial wall, and that could be disastrous. This new laser will make the ones being used now seem like sledgehammers."

Using fiber optics in the body is not new, however. As long as twenty years and more ago, doctors were slipping fiber optics down the throats of patients to search for swallowed coins and elusive tumors, to push all the way to the duodenum, to reveal the machinery of the body as nothing else can. Even before the development of fiber optics suitable for relaying communications over long distances, the endoscope and gastroscope were being used. Composed of several bundles of fiber optics, the instruments enter painlessly into the body through the mouth or nose, or even the rectum, to send back images for display on a screen. The patient is usually semisedated, conscious enough to wonder why those in attendance seem to be paying prolonged attention to one of the images. Such scopes, with their light-carrying glasses of depthless transparency, lift diagnostic procedures to new levels of success by detecting many cancers not visible on X rays. Also, by using instruments carried by the scopes, surgeons can make certain corrections as they scan, such as removing a polyp from the colon.

In medical laboratories around the world, glass has been the dominant material in beakers and vials, flasks and petri dishes, and most other labware. The indispensability of glass for the medical technician has been ensured down through the years by the material's resistance to erosion by chemicals and its inability to contaminate samples. The transparency of glass is vital, as is the way it lends itself to easy sterilization. Still, plastic is making inroads in the lab; indeed, the mother ship of all labware, the common beaker, is now available not only in plastic, but in paper and platinum as well. Glass breaks, and if it doesn't, it has to be sterilized again. Plastic clears both of those hurdles. So it is here now—the disposable pipette.

New plastics of good transparency and with optical qualities—the ability to refract light and pay out color—are coming on the scene and before too long glass may be pushed aside, never to regain its dominance. For now, however, the material continues to hold a strong presence in the labs, especially the borosilicates glasses, the types of glass of which Pyrex and Schott's Duran are made.

For the most part, glass labware is mass-produced, and the designs of the pieces remain tedious in extreme. The beakers and vials have been aesthetically bankrupt forever, and as for the small bottles meant to hold pills or samples of bodily fluid, they, clearly, were not what Robert Browning had in mind when he wrote in "The Flight of a Duchess" about a vessel made of Venetian glass, a jar "with long white threads distinct inside, / Like the lake-flower's fibrous roots that dangle / Loose such a length and never tangle."

Special pieces continue to be blown by hand, such as a replica of a body part to be used as a teaching tool in medical school. There are between six hundred and a thousand scientific glassblowers in the United States today, and they make many of the specialty pieces needed by surgeons and others in the medical field. Some of those craftsmen work freelance; some are employed by industries, others by universities.

Richard Grant of Dayton, Ohio, worked thirty-seven years as a scientific glassblower, and he does not like to be confused with the person who entertains at a store opening by crafting birds in glass. "My work is associated with science, not animals," he said with not so much as a hint of pomposity. "It is a craft, not an art." He was at the University of

Dayton for most of his career, and in his work there he generally made one-of-a-kind pieces not available through the catalogs of vendors. If a scientist had only a cloudy idea of the type of instrument needed, Grant would assist in the planning.

"Glass works very well as a material for a prototype," Grant said. "It's easily changed to get exactly what is needed. I'm not very artistic, like a lot of glassblowers, but I am a perfectionist." The demand on those who craft scientific pieces in glass, using a gas-fired torch in what is known as flameworking, is not so much creativity as it is perfection. It couldn't be any other way when Grant was enlisted to fabricate a device that could extend the life of a patient who is near death with a failing heart. It is called a heart assister, or, more formally, direct mechanical ventricular assistance (DMVA). It is round and rather deep, shaped like a goblet but suitable for serving *pêche melba*. Except for an inner lining of a latexlike diaphragm, the device is all glass, a good, strong glass, a piece meant to cup the heart and stroke it to raise another hour—a day, even, or more—of life.

A fellow townsman of Grant, Dr. George L. Anstadt, a veterinarian, invented the DMVA in the 1960s, but it took the medical community about thirty years to acknowledge the important potential of the instrument. In 1990, a fifty-six-year-old woman was admitted to Duke University Medical Center in Durham, North Carolina, facing imminent death because of heart disease. Surgeons opened her chest and attached the glass to the enlarged organ so that it covered the lower chambers. Then air was pumped into the unit, causing the diaphragmatic lining to swell and deflate, massaging the heart as a doctor would do manually as a last resort to keep a heart patient alive. The glass device, crafted by Richard Grant, stayed on the woman's heart for almost two and a half days, until she received a transplant.

Dr. Anstadt says there will never be a complete artificial heart, but he does believe that his cup can remain in the body for six months, or even a year, to provide some insurance for the continued circulation of blood in the event of a cardiac episode. "This device of mine has been around a long time," he said. "It may have to be reinvented to gain the attention it deserves." Meanwhile, we have our memories of a glass body part that was in use for several centuries (and still is, in some places), a glass prosthesis that offered nothing physiologically but was a psychological benediction. For the disfigured, the glass eye was a slayer of despair.

The social history of the glass eye is enriched by the unrivaled value placed on the human eye. To lose one's sight is generally regarded as a tragedy that pushes death for depth of loss, and to have an enucleated socket to show for it is all the more cause for sorrow. It is not known when artificial eyes were first used. But if the ancient Egyptians honored their dead with jeweled inlays of representations of eyes in mummies and sarcophagi, would they not have fashioned a false eye for use by those still alive?

The glass eye was probably introduced in Europe sometime during the Renaissance, in time, it can be imagined, for Shakespeare to have given these words to King Lear: "Get thee glass eyes; / and, like a scurvy politician, seem / To see things thou dost not." Early glass eyes were made by working a hollow glass rod with the heat from the flame of a lamp, getting a round melt on the end of the rod and fashioning that into the shape of a ball on which the iris colors were painted. The pupil was black glass. Sometimes the prosthesis was affixed to a leather shield, on which the eyelids and brow would be painted. The whole piece would be held in place on the face with the use of a band that went around the head.

Facial prostheses in the early years of their making were,

on the whole, lacking in cosmetic refinement. Sometimes the artificial eye was part of a full rubber mask. At other times, the eye might be fused to the metal rim of eyeglasses and held in place on the face in that way. In the case of noses, glass, fortunately, was passed over for metal, including the silver used to make noses for Turkish troops mutilated by Russians during fighting in the 1870s.

Metal was also a popular material for artificial eyes in early times, but by the middle of the eighteenth century, it became widely known that glass was more biocompatible. It was even said that tears lingered on glass longer than on metal, and that lent a quality of sparkle to the false eye. By the middle of the nineteenth century, French oculists were considered the finest in the world, but it was a German, Ludwig Müller-Uri, who took the glass eye to a new level of perfection. Discarding the use of lead oxide glass, he introduced a new hard, light glass made with cryolite and arsenic oxide. In that, sodium aluminum fluoride is present to give the eye a color closely resembling the gray-white of the sclera.

Itinerant oculists from Germany began to arrive in the United States, and by the turn of the twentieth century, the business of making glass eyes was firmly established in the land. The glass was imported from Germany, and with the passage of years, the quality of the work improved to where the prosthesis, artful now, if still unseeing, could lead the wearer across the barrier of fear and repulsion that separates the deformed person from those who lack understanding.

With the onset of World War II, Germany embargoed shipments of the special glass to the United States. Meanwhile, casualties involving loss of eyes were mounting among the forces of the Allies. Commissions were appointed to find answers. They agreed to make artificial eyes out of acrylic plastic, and to this day most still are, except in Germany, of course, where glass is still preferred.

Glass has so many forms, so many different formulas, that there is usually something there to fill the need for a material in a medical procedure. The work comes in finding the right glass. Now, a new glass has been found: a glass-ceramic, something unlike glass as we know it. Glass as a mover of radiation to a tumor in the liver? Yes, it makes sense. Glass as a device to massage the heart to ward off death? Yes again.

But glass as teeth?

Chapter Eight

With the Toughness of Steel

High technology has spilled over into the dentist's office, there to set off small bursts of change in dental restoration work. Glass is being brought in as a material for saving a tooth, or, as is being done in Europe, for making full dental prostheses—teeth, roots, and all. And using glass, the work is done in one sitting, and without the need for having impression-taking wax—warm and gummy, like Play-Doh that's been squeezed in the offensively moist hands of a feverish child—stuffed in your mouth.

The material belongs to a family of glasses invented more than forty years ago, but only now, with refinements, coming in for wide use. It is rightfully a glass-ceramic, a material of partial crystallization so tough it can be worked on a lathe, making a machinist of the dentist while he's creating a replacement part for a tooth. First, a picture of the tooth is taken by an intraoral camera. With the image transferred to a screen, the restoration of it can be designed by use of a computer, and when that is done, a block of the glass-ceramic of a color to match the patient's teeth is inserted into the milling chamber of the lathe; it takes about five minutes to craft the piece. The restoration is etched in acid, applied to the tooth with a bonding agent, and given a final contouring and polishing.

All of that is done in a one-hour session with the dentist, and while the time is brief, there is no shortchanging in quality. The glass-ceramic is stronger than dental porcelain, plaque-resistant, and highly translucent. The work is all done with one compact unit, called the CEREC, a piece five feet tall at the most, and two feet wide and deep. Everything from lathe to camera is in that one small unit.

"More and more people are keeping their teeth now," said Dr. Stuart Ross of Washington, D.C., president of the Association of Computerized Dentistry. "People are saying, 'Will you please restore my tooth?' rather than 'Will you please make me a denture?' " In the old days, that meant putting in a filling or a crown, but with the computer and this glass-ceramic, a whole new field of possible treatments has opened up. Rather than a full crown, a dentist can now reconstruct a tooth with this new, advanced technology by having the computer design, and the lathe mill, inlays and onlays and even cusp replacements. Discoloration can be corrected, and tooth fractures repaired. There is no waiting period during which time the patient must make do with uncomfortable and unattractive temporary props in the mouth.

There are about 300 CEREC machines being used by dentists in the United States, and of the only two in Washington, D.C., Dr. Ross owns both. "The machine costs seventy thousand dollars," he said, "and that's a lot of money for a dentist. But over a period of five years, it pays for itself."

Ross, who is forty-seven, agreed that older men and women of the drill are reluctant to break out of that ageless mold of single-dimension dentistry. Drillings and fillings and "open wider, please" exhortations continue to drive the technology of general practice; restorative work in the back of the mouth is still aglitter with heavy deposits of gold and amalgams that incorporate mercury. (If a mercury spill is considered a dire threat to the environment, why is the poisonous element used in the mouth? Because, says the American Dental Association, dentists have been using mercury in fillings for 170 years without ill effect, and there is no reason to ban it. Not so, says the International Academy of Oral Medicine and Toxicology; there is reasonable doubt about the safety of dental amalgam made with mercury. The federal government does not seem inclined to come down on one side or the other, and so the issue is likely to continue sparking, as it has for many years.)

Until I moved in the mid-1990s, my longtime dentist was a man named Irwin, who had a driving compulsion to attend auctions and bid on antique toys. He insisted he was not personally put off by computerized dentistry, and he seemed intrigued with the concept of using glass for tooth reconstruction. But his patients, he insisted, had come to like the old-shoe feel of his office and, indeed, of Irwin himself. "I have to give them what they want," he would say. "And glass for teeth? That might be a hard sell."

A glass-ceramic is a glass on which crystals are made to grow. In the beginning of the process, a nucleating agent with a high tolerance for heat, such as titanium dioxide, is added to the batch for melt. Growth of the crystals (called "crystallites" at birth) around the nuclei—one crystal for each nucleus—is controlled by temperature. Crystallization takes place with heating and reheating, and in the end, the crystallinity can account for almost all of the material, if that is desired. Each type of glass-ceramic is defined by its glass content and by the type of crystals it contains. But on the whole, the material has an honors list of properties, such as its high immunity to the shock of heat, hardness, and a strength four to five times that of standard glass. A block of glass-ceramic just out of the oven will not burn the fingers when held by the corners, although the center of the piece may have a temperature of 2,200° F.

With the heat confined in that way, and with the lack of warping, glass-ceramic material serves nobly in the kitchen, not only as vessels for cooking, but also as the cooktops of ranges, where the induction heat is fueled by either gas or electricity. The pot is directly on the glass-ceramic, in a position to be simply pushed onto a surrounding cold zone when the cooking is done. No longer will your spaghetti sauce, when boiling over, escape into the canyon of the burner; rather, it will wait for you and your rag on the range top. Away from the kitchen, glass-ceramics, with excellent electrical properties, have been used in the electronic systems of guided missiles, and, more to the deadly point, as the housing for the radar antenna on the nose of the weapon.

The glass-ceramic is a Corning discovery, something that Dr. S. Donald Stookey, a chemist at the company, came upon when, as the story goes, he placed a piece of glass in the laboratory furnace, set the temperature, and left the room. When he returned, he found that the heat in the furnace was 600° higher than the control setting. The glass was now cooked to discoloration, and when Stookey removed it from the furnace, it slipped, fell to the floor, and, with a clank alien to glass, bounced. He realized then that the temperature change had crystallized the glass and transformed it into a highly useful ceramic material tough enough to withstand heavy impact. In a tribute to Stookey, Corning has memorialized him as one not to let a mistake pass by without first frisking it for some possible good. (Actually, a lot of the major discoveries and advancements having to do with glass came about as a result of something that went wrong.)

Stookey made his discovery in 1957. Since that time, glass-ceramics have been nudging metals aside and becoming the material of choice for a number of applications. One potential use that seemed particularly promising was as the substrate for a magnetic memory disk—the hard disk on which to store material in a computer. Aluminum alloy has dominated that market from the start, but there are problems with disks made of metal. They must be thin, for one thing, sometimes no more than half a millimeter. And when aluminum is reduced to that thinness, it starts to act like foil.

Using a parent glass similar in composition to the common soda-lime glass except for the presence of fluorine and a small amount of alumina, Corning developed a glass-ceramic that was ideal for use as a substrate for a magnetic disk. They started with what looked like plain window glass, and treated it with 1,650° F heat until the crystals rose like seedlings—long, finger-shaped crystals tightly interlocked. Seen under high magnification, the crystallized glass flashed with its toughness. "With a glass-ceramic like this, a fracture would sort of go all around the crystal and through the cleavage plains," said Dr. Linda Pinckney, a Corning research associate who worked on the early development of the material. "And lots of times, a fracture will go part way through and just stop. At those times, it won't break at all. It is a very tough material. You can throw a disk made of this glass-ceramic around and it won't dent like aluminum." The resistance a glass-ceramic puts up against surface abrasions is what gives it outstanding qualities of toughness.

When glass is made to grow crystals, it sags and flattens out, and flatness is important for the successful functioning of the hard disk that a computer writes to and reads from. When the disk starts to spin, the head flies just above the surface. It is desirable to have the head as close as possible to the surface because in that way it can see more and read more, and pack data more densely. But the disk is spinning at 3,600 rpm, and if it is not perfectly flat and rigid, or if there is anything on the surface of the disk that doesn't belong there, there is likely to be a collision with the head. The result never fails to terrify: a computer crash.

In 1993 Corning was making the disk substrates at a plant in Danville, Virginia. On the surface, at least, the market for fixed disks for use in computers was secure. But once again, the company, exhilarated by the research chase, seemed to lose interest with the captive product. Two years later, Corning stopped making glass-ceramic disks. When asked why, company officials gave answers that were splintered with vagueness, but essentially they were saying that profit potential, without undue delay, was the determining factor. Glass that is not crystallized but has been strengthened by chemical hardening is now being used to make the disks in Japan. It too is strong, but glass-ceramic, with its crystal armor and its ultraflat thinness, carries less of a risk factor.

We're at a time now when it's possible to walk into a machine shop, as I did one day, and find workers fabricating glass-ceramic nuts and bolts. It is glass with a machinability that is precise, glass that is being sawed and punched and drilled. It is being used for seals and gauges, measuring rods, blanks for telescope mirrors, and other optics of all kinds, and for materials in advanced navigation technology, such as laser gyroscopes. Hundreds of parts made of glass-ceramic can be found on the space shuttle.

Glass-ceramic is a superb insulator at high voltages. A company in need of a device to handle fifty thousand volts, and not be larger than two inches by six inches, is not likely to have suppliers stampeding to the door. "To get a piece of insulator that size and for that voltage was just not simple," said an official of one such company. "We did it by having a glass-ceramic machined down to fit the system, and then we put it together. Bingo."

No doubt the largest piece of glass-ceramic fabricated so far was done at Schott Glaswerke in Germany, where the first of four mirror blanks being made will be part of the world's largest optical telescope.

Glass-ceramics have advanced the quest for strengthened glass by a large step. Usually, the material doesn't fail when under compression, and it is compression, of course, that gives it strength. Few things loom so challenging for glass scientists than the great unknown that spans the theoretical and practical strengths of glass, and with glass-ceramics the two extremes are being brought closer together. The problem remains, however, that the atomic structure of glass does not allow the material to deform at room temperature. When a load becomes too much for glass to bear, it simply fails, and not in a docile manner. There are no partial fractures with plain glass.

Glass-ceramics are strong—very strong—and so are some chemically treated glasses, but if the ductility of molten glass could be at least partially extended into its solid state, that would be a glass of truly superior bending strength. If glass is going to deform in order to carry large stresses, it needs help. It needs something like fiber. A piece of bamboo, for example, does not break cleanly. It bends and eventually cracks, but the plant's fibers hold fractured bamboo together as it fails gracefully.

Fiber-reinforced glasses are being made today, and they offer a bending strength that is more than ten times that of nonreinforced glass. One way of making them is to bind bundles of the fibers with a powder of finely ground glass, and put them through a process involving heat and pressure.

When this glass breaks, it does so not with the rifleshot finality of brittle glass, but more with the compromising stretches and bends of bamboo.

The fibers are made of carbon or silicon carbide. The composite is usually 60 percent glass and the rest fiber. "This glass has enormous potential as a construction material," a technician at Schott said while dropping a steel ball from a height of four feet on a piece of the still-intact reinforced glass. "It has a big advantage over steel in that fiber-reinforced glass is so much lighter." It would be very suitable for use in airplanes, and in automobile engines as rings for pumps and other parts. Being glass, as a nonconductor of heat, this could eliminate the need for some of the cooling and lubrication in engines. Because of its immunity to thermal shock, fiber-reinforced glass has strong potential as a material for tools used in hot shops—in shops where glass is made—and, indeed, it has been put to use in that regard.

Glass reinforced with carbon fibers is black, and while that may not be suitable for use in the window of a cathedral, it does give emphasis to the uncommon strength of the material. It is a glass with a rhinoceroslike stamp.

The glass that meets the test for the most rigorous requirements of resistance is, for the most part, supertough laminated glass, incorporating the use of plastic. The most common use of laminated glass is, of course, for the windshields of cars. To make it, two sheets of glass are laminated on either side of a layer of viscous plastic, such as polyvinyl butyral resin (PVB). Should the windshield break, the shattered glass adheres to the plastic rather than flying off in deadly shards, and that safety feature is responsible for prevention of many thousands of deaths and serious injuries. Safety glass is the best of concepts—simple, and yet hugely rewarding in the saving of lives and prevention of injury.

It takes more than a thin windshield laminate to stop a

bullet traveling 2,400 feet per second, or more than twice the speed of sound. A round from a 30-60 rifle moves that fast, and when it strikes the laminate of glass and plastic, the bullet will be stopped even before the sound of its firing catches up. Again, as with a windshield, the glass cladding will shatter but not splinter. But in this case the glass would need to be at least 2 inches thick, as compared to the standard 1³/₁₆-inch shield used to protect bank tellers.

Laminated glass is often used in the construction of airports because of its sound-control qualities, and another popular use is for large conservatories in botanical gardens. But as a resister standing against threats to life and property the glass has gained recognition beyond that of being the favored material for windshields. In the end, as generally perceived, the bullets, the shearing winds of hurricanes, the bricks of rioters are what truly test the strength and toughness of a glass.

And the attacks of the deranged with their bombs and hammers and knives on buildings, and on masterpieces of art in the museums of the world—they too are tests. In late July 1993 a bomb planted in a car exploded outside of the Uffizi Gallery in Florence, causing deaths and destruction enough to seemingly bring tears to the allegorical woman with flowers of Botticelli's *Spring*. That painting by the artist, along with his *Birth of Venus*, and Michelangelo's *Holy Family*, painted in 1450, were spared because they were protected by laminated glass. Twenty other paintings in that greatest collection of Italian art were damaged, three of them beyond repair.

On the day after the bombing, twenty thousand people joined in a procession through the streets of Florence. Theirs was a collective wounded pride, a sense of betrayal to their heritage, and no glass shield exists to protect anyone from that.

Tourists in Italy, generally, are critical of protective glass for art in museums there. They can no longer get close to Michelangelo's *Pietà* in St. Peter's Basilica because it has been attacked, and there are paintings, to which the craquelure lends authority of age, held off from the crowds by glass. Of course, sculptures by Bernini fill the Piazza Novona in Rome, and they can be climbed and sat on. And in the rear of Rome's San Pietro in Vincoli (St. Peter in Chains) Church, cast in a sorrowful light, is Michelangelo's powerful sculpture of Moses, *sans* glass shield; stand close to it—you're allowed to do that—and the genius of the work will race through you like an electrical charge.

Moses may be out in the open and free of protection, save for a few observant priests walking patrols in their thick-soled black shoes, but the pope must be in a box of glass when he moves along the roads of the many foreign lands he visits. He is in his Popemobile, a transport brought into service following an attempt by a Turkish gunman on his life in 1981. It has none of the sleek lines and speed of a Porsche, and there are those among the devout who say it is not a proper vehicle at all for a pope to be riding in. But it goes where he goes—to Rwanda and Sarajevo, and Cuba, and other places in the grip of despair—and that is how we have come to envision Pope John Paul II, standing, gesturing, and willing his blessings on the people through a bullet-stopping laminated glass.

The glazing technology for making laminated glass with PVB, like that used for the Popemobile, was developed by E. I. du Pont de Nemours and Company. The company first produced the polyvinyl as a laminate with glass in the 1930s, but rather than security, it was used at first for automobiles. Architects used it, too, and because of it the Bank of America building in Los Angeles survived an onslaught of Molotov cocktails and bricks during a riot in 1992. The mobs broke but did not penetrate that glass.

The use of laminated glass as an effective barrier against the massive damage and high death count of some terrorist attacks drew considerable attention in the mid-1990s, following bombings of the World Trade Center in New York, the Federal Building in Oklahoma City, and a housing project for American soldiers in Saudi Arabia. In Washington, fearing for the safety of President Clinton and his family in the White House, the Secret Service closed off Pennsylvania Avenue in front of the mansion in a move that has been described as "bunkerism." Closing that stretch of the avenue between the front of the White House and the southern side of Lafayette Square drained much of the spirit of nationhood that resided on that block.

"The street is sort of still there—it's still all paved and so forth—but it's empty except for a skateboarder going by now and then," said Arthur Cotton Moore. "It's very strange. With the barriers up, it has a sense of being like the Berlin Wall."

Moore is a well-known Washington architect, an innovator and designer of high regard. He, like many others, is appalled that the White House has been closed off from the people. He is also aware that there is reason to be concerned about security, but closing Pennsylvania Avenue in front of the White House amounts to, in his words, "a victory for the forces of lawlessness." He has proposed another way to counter the threats of bombings by terrorists: Erect a fence of laminated glass across the front of the White House, just behind the metal fence that has long been in place.

The distance from the edge of Lafayette Square, across Pennsylvania, and up to the front of the White House is four hundred feet, give or take a few feet. "Tests have shown that the explosion of a five-hundred-pound bomb that had been placed as close as possible to the laminated glass fence would only crack the glass, not destroy it," Moore said. The Secret Service, adamant on keeping the avenue closed although Congress has voted overwhelmingly to reopen it, maintains that a five-thousand-pound bomb detonated in a truck on Pennsylvania Avenue in front of the White House would be disastrous, resulting in deaths and structural damage to the building. Experts, however, dispute that, pointing out that laminated glass performed exceedingly well in the Oklahoma City explosion, having survived while every other window was destroyed.

The glass fence that Moore proposes would be between one and a half and two inches thick. For extra precaution, he said, a second fence of thinner glass could be added at a distance of about three feet behind the first one, with the space between the two planted with shrubbery. The cost: an estimated $2.5 million to $3 million. The Secret Service, in effect, would like to have Lafayette Square, which stands in front of the White House, extended to cover the section of Pennsylvania Avenue now closed out of fear of a terrorist attack. That would cost over $40 million.

It isn't as if laminated glass hasn't been tested in the service of the United States. The glass has been protecting the Liberty Bell in Philadelphia since 1987, when a brick crashed through the window of ordinary glass in the pavilion where it is displayed, and as much as sixteen thousand pounds of the material is in use at the Statue of Liberty in New York, protecting both the statue and those who visit it.

But to get an appreciation of this glass on a human level, seek out a small liquor store in a neighborhood where clipped English boxwood does not grow—preferably a liquor store offering check-cashing services as well. By better than fifty-to-one odds, there will be a clerk there encamped behind enough laminated glass to stop stampeding buffalo.

Such glass can embolden a person otherwise faint of heart to stare down a .44 Magnum about to let go with a 240-grain soft-point bullet traveling at 1,470 feet per second.

Glass does not perform as well when challenging the violence of a hurricane or tornado, but only because of weak building codes, or laxity in enforcing stronger ones. In south Florida, it is a ritual to tape windows or cover them with plywood when a tropical storm is swelling in force as it heads for landfall, for the glass in those windows, for the most part, is not structured to withstand raging winds and debris flying at missile speed. Windows reinforced with vinyl, on the other hand, may break, but the glass will hold together.

Early in the morning of August 24,1992, a hurricane came ashore on the tip of Florida and began an onslaught of destruction that would become recognized as the worst natural disaster in the history of the United States since the San Francisco earthquake of 1906. To this day, the name Andrew draws on the apprehension and fears of Floridians during the three months when such storms are born—and die trailing death. Some eighty-five thousand housing units and buildings were destroyed, and six years later you can still find broken glass along the trail wrenched out by Andrew's winds of more than 150 miles an hour.

In a bombing such as the one that destroyed the Federal Building in Oklahoma City, a condition called spalling usually occurs; that is when ordinary window glass is shattered by flying debris with such violence that the back surface of it explodes, launching shards that can injure and kill. At other times, the entire window can be blown out of its frame. In either case, a hole is left in the structure, and if the force is that of a hurricane, the winds and rains can push through just one such opening to destroy a house.

It was debris, though—specifically heavy roof tiles, not

glass—that caused much of the extensive damage—up to $20 billion in property values—during Andrew's excursion ashore. Roofs in the storm's path disintegrated to an extent beyond the force of the winds alone; the quality of construction, it was later revealed, was in gross violation of even a weak building code. It was like a revelation of good and evil to walk along certain streets in Miami's Dade County and find all the houses on one side leveled or heavily damaged, and those on the other still standing with roofs in place. The reason, of course, was different builders.

"We had not had a major hurricane in Florida since 1964, and we had become rather complacent as a community," said Ira D. Giller, a Miami Beach architect. "That all changed after Andrew. There have been major changes in the building code. For glass, the changes involve surging center protection, and resistance." That means, in this post-Andrew era of construction in south Florida, that the center of any glass in a window now must bow and spring back when struck by debris, as the strings of a tennis racket do when meeting a 100-plus-miles-an-hour service.

As for resistance, more laminated glass is now being used for construction in the hurricane zone. When Ira Giller was the architect for transforming an old synagogue on Miami Beach into the Stanford L. Ziff Jewish Museum of Florida, he incorporated the use of a du Pont glass composite called Sentry-Glas to protect windows of stained glass. Constructed of annealed glass, PVB-glass laminate, and polyester, the composite puts up a formidable resistance against vandals as well as wind and debris. In its first test, the intrusion-resistant laminate prevented any damage being done to the stained glass when bricks were hurled at it from the outside.

Sometimes now in late summer, when all the tribes of violent tropical weather are doing war dances in the Caribbean and Catholics are offering prayers to Our Lady of

Prompt Succor to protect them from the winds and rain, the thoughts of a lot of people in Dade County and other southern coastal areas turn back to that morning in 1992, and they speak of it now not so much with fear as with awe. They know that another storm of such gorilla strength can come at any time, and there is interest in how buildings erected under the new and strengthened codes will fare.

At the Hurricane Test Laboratory, Inc., in the Florida east coast town of Riviera Beach, mockups of new window systems are being tested to determine if they meet the standards set down in the revised building codes for coastal communities in south Florida's Dade, Broward, and Palm Beach counties. "Under the new codes, a window must have hurricane shutters, or it must be designed to withstand the impact of flying debris," said Vinu Abraham, general manager of the laboratory. "Shutters are very effective, but there is one problem with them: You have to put them up when a storm is approaching. With laminated glass in the windows, that's in place all the time."

The test for impact tolerance is this: A four-by-four wooden beam, seven to nine feet long, is fired from an air cannon at a velocity of thirty-four miles per hour. It must not penetrate the glass or open a hole in it. All of the glasses being tested are laminates, for they are the only ones of window-grade clarity capable of taking such a hit without penetration.

Today, strengthened glasses with clarity of high quality and aesthetic integrity are claiming more and more space in architectural renderings. No longer is laminated glass the unglamorous Supphose of the material. Significantly refined, now it suits chapels as well as cashiers' cages. It fronts the high-priced skyboxes at sports arenas (the wealthy need to be protected from foul balls and the rowdy actions of bleacherites). It is used for large atriums and for the canopies of sunlight-flooded malls, including the Mall of America in Minneapolis, the world's largest; for that massive bivouac of shops on the prairie, they covered a six-acre area with 250,000 square feet of laminated glass.

More than just for protection, laminated glass is valued by architects for its design flexibility as well as its acoustical and solar control, a factor in the decision to make extensive use of the material—660,000 square feet—as the exterior curtain for the tallest building in the world, the Kuala Lumpur City Center in Malaysia. At 1,483 feet tall, the twin towers of the City Center stand thirty-three feet higher than the Sears Tower in Chicago. There is enough of the glass in that building to cover the streets of fifteen city blocks.

And it is something—to be able to be high up on, oh, the seventy-fifth floor, and look out beyond the city, out to distant jungle reaches of the Malay Peninsula, the Golden Chersonese. It will put you in mind of a magic world where wild elephants foray for the fruits of the oil palm, and diamonds and gold set sultanates aglitter.

If laminated glass goes high up, as in a skyscraper, then hollow glass goes way down, as on the floor of the ocean. Hollow spheres of this glass are buoyant and ideal for use at depths of twenty thousand feet or more, down where the pressure can collapse steel. Another form of glass is made of nothing other than silicon oxide. Called fused silica glass, it is one of the purest substances ever manufactured. The melt for this highly specialized (and very expensive) glass must be at a temperature in excess of 3,600° F. It is more resistant to thermal shock and chemical attack than any other glass being made.

Among the limited uses for fused silica glass was for the outside pane of the double windows on the lunar excursion module that tootled on the surface of the moon. Indeed, fused silica has been used for the windows of all manned space vehicles. This glass is ever-increasingly used in optical applications, for optical glass, being wed to light as no other material can ever be, deserves the best.

Chapter Nine

Raising the Invisible to Sight

Of all the glasses, none to my mind is so noble as those that gather in light and usher it through, but sometimes bending it, focusing it, bouncing it back, reflecting or refracting it, taking even dim glows in the night sky and turning them into torches revealing the dark and distant mysteries of uncharted space.

Optical glasses were first made in the late eighteenth century. From the start, they have been high-quality glasses free of impurities, glasses ground to precision for use as prisms, and as mirrors and lenses in telescopes and microscopes, cameras and binoculars. The performances of optical glasses have been extraordinary, especially in meeting the needs in this age of space exploration for advanced precision instruments. For example, as a lens in a camera positioned on a space probe twenty miles above the earth, it allows light to pass with little loss and that lets the camera get clear pictures even under poor lighting conditions; and with that glass in that lens, at that altitude, it is possible to get pictures in which landmarks as small as individual buildings can be identified.

The French were the first to make optical glass of good quality, and they did it by stirring molten glass in a pot with a piece of clay. The glass was allowed to cool slowly in the pots of three-foot diameter until it cracked into large blocks, after which it was chipped to get at and remove the "stones"—bits of unmelted batch and pot refractory in the blocks. Then the glass was reheated and given shape as lenses. This was glassmaking by attrition, in that they started out with a ton of glass and ended with just 200 pounds of optical lenses. Because metal, except for aluminum, will dissolve in the intense heat of the glass melt, clay, or ceramic, pots continue to be used, although vastly improved now.

Since that time, the pot has been largely abandoned in favor of crucibles made of platinum. Unlike before, very small pieces can be ground and polished. Indeed, there are spherical lenses of high-quality optical glass being made today that are no larger than the tip of a pencil; they are used in players for both audio and video disks. On the other end of the scale, there are telescope mirrors being cast that are epic in size—disks weighing twenty tons and taking up two lanes of a highway when moved by truck. Optical glass: romancing light large and small.

As a pioneer in the field, Otto Schott of Germany had developed 44 different optical glasses by 1886 and by that time he had established a working relationship with the famous precision engineer of optics, Carl Zeiss. Schott glass was used for the first Zeiss photographic lenses, produced in 1890. It took Schott nearly a hundred trial melts to obtain a glass of suitable optical qualities.

Of all the optical-glass items made by Schott, none matches in size the ones now being prepared for the telescope that will be placed atop 8,740-foot-high Cerro Paranal in the Atacama Desert of Chile, a site where most nights are awash in brilliant clarity. The instrument is being made

The Corning Museum of Glass

Head of Amenhotep II, 1436–1411 B.C., 18th Dynasty

The Corning Museum of Glass, Gift of Arthur A. Houghton, Jr.

The Morgan Cup, first half of the 1st century A.D., Roman Empire

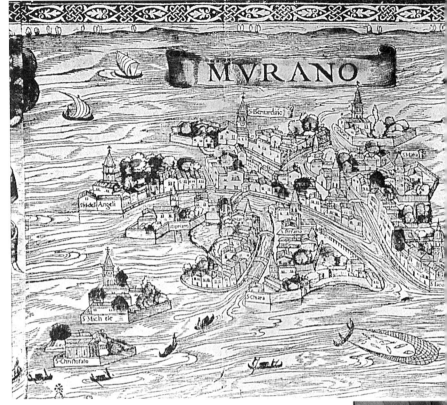

A view of Murano. Published by Matteo Pagan.

Aquatint after J.M.Volz by C. Meichelt. From Glashutte im Aule (Early 19th Century Glassmaking in the Black Forest) *by A. Schreiber, Freiburg, 1820–1827.*

The Corning Museum of Glass

Collection of the Juliette K. and Leonard S. Rakow
Research Library of The Corning Museum of Glass

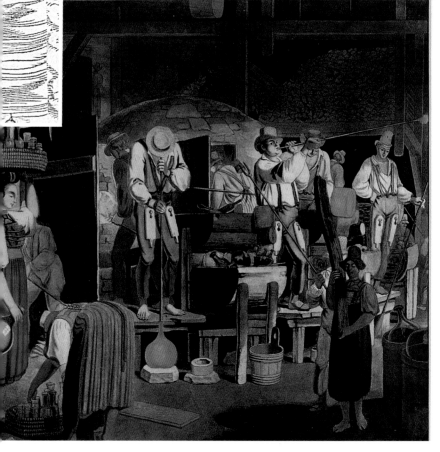

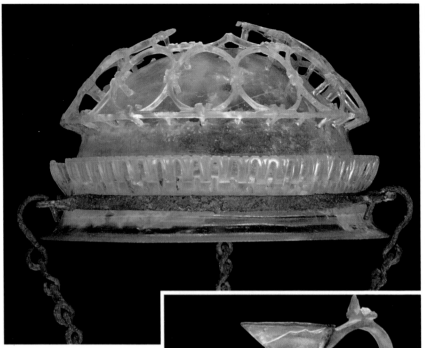

The Corning Museum of Glass, Gift, funds from Arthur Rubloff Residuary Trust

Cage Cup, about 300 A.D.,
Roman Empire

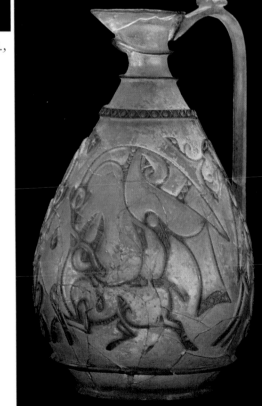

The Corning Ewer,
late 19th century A.D.,
probably Northern Iran

The Corning Museum of Glass

Tiffany landscape window from Rochroanne, Irvington-on-the-Hudson, New York, 1905

Corning Incorporated

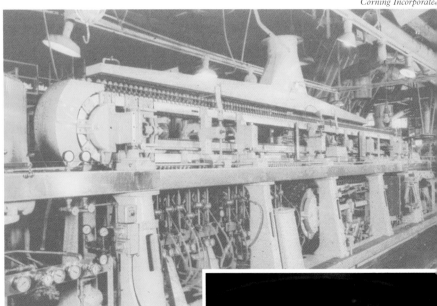

Ribbon machine in Wellsboro, Pennsylvania

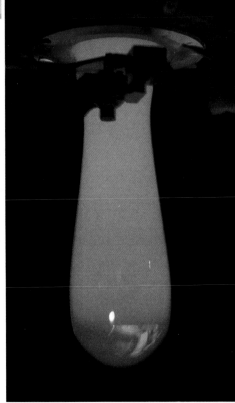

Gob of glass used in the manufacture of large screen TVs

Corning Incorporated

Corning Incorporated

Fiber optics with color laser tips

Telescope mirror blank

Chiostro di Sant'Apollonia, turquoise, Finland, by Dale Chihuly

Rob Whitwe

Translucent Yellow Seaform Persian Set (detail), 1994, by Dale Chihuly

for the European Southern Observatory, a consortium of eight European Community nations. It will be called the VLT, which stands for (no acronymic pretentiousness here) Very Large Telescope. Each mirror, made of a glass-ceramic called Zerodur, will be nearly twenty-seven feet in diameter and weigh twenty-three tons. It has been said that this will be one of the most stupendous works in glass ever done.

"Ever done, yes," agreed Alfred Jacobsen, a Schott official and an expert on optical glass. "But you must understand that it is very difficult to handle such a large piece, and it must be transported a long way."

"Are you afraid it's going to . . . ?" (It is difficult to speak of breakage to those who work with glass for a living.)

"There *is* concern," Mr. Jacobsen replied, going on to point out that while one of the mirrors can cover the floor space of a double garage, it is less than seven inches thick— in essence, a wafer.

The melting tank alone was as tall as a two-story house, and on this day it held seventy-five tons of molten glass. The shaping was by a spin-casting process, and as the molten glass was released into a mold, which has been preheated for two days, the heat rose and the glow of the fiery glass spilled through the shop. It took more than three hours to pour forty-five tons of the glass, holding it at a temperature of 2,500°F, into the mold. The spin cycle was set for six revolutions a minute, enough to hurl the glass into the desired contours.

Then came a phase in the casting when intimations of faintheartedness were in the air: The mirror had to be removed from the mold. The workers were silent in concentration and hardly noticed that the temperature in the room had reached 122°F. A seventy-ton crane moved into position above the mold, and from it dropped lines attached to vacuum cups. While held by the cups, as the mold was dismantled, the mirror hovered there like a phoenix rising from its natal fires.

It took four months for the mirror blank to cool to room temperature, and at that point it was still pure glass. To transform it into a glass-ceramic, workers at Schott put the piece through a series of heat treatments over a period of a year. Finally, with the meniscus blank now 70 percent crystalline, it was crated, resting on air springs (no Styrofoam peanuts used), and trucked to the banks of the Rhine River in Mainz for travel to St. Pierre–du–Perray, near Paris, for two years of polishing.

The mirror was positioned on 150 hydraulic tripods for the polishing. There a robot controlled the wooden disk that did the polishing by rubbing the mirror with a fine grinding agent called Ceroxid. God forbid that the grinding disk ever faltered or the mirror wobbled, for this work was like a ritual in a culture of nanometers. All lines and angles were true; accuracy tolerances were next to immeasurable at most. The first phase of polishing lasted for six months, and when that was completed, the mirror blank was mapped by a laser to determine what corrections were needed. Such periodic testing continued through the full polishing until, in the end, the mirror was pronounced to be within a few ten-billionths of a meter of perfection.

All four blanks have been cast by now. What remains to be done is to get them from Paris to Chile by ship and then, knock wood, up to the top of the mountain in a specially made vehicle with sixteen axles. Throughout it all there was a sense of mission in the casting hall at Mainz, as if the workers shared an awareness that when the four massive mirrors are in place in the clear air of that high desert in Chile, where the land is the driest in the world, they will reach out beyond the stars to see where we've never seen before.

Corning made a telescope mirror blank larger than the Schott one, but rather than being of a single cast, it consists of many hexagon-shaped segments fused together. The hexagons themselves were made of fused pieces called boules. Weighing thirty-five tons and measuring twenty-seven feet in diameter, the blank was made in Corning's Canton, New York, plant for the Japanese government, and is expected to be in operation on the summit of Hawaii's 14,000-foot-high extinct volcano, Mauna Kea, by the year 2000. Like the Schott mirrors, this too was barged to France for three years of polishing. When placed in its housing, the mirror will be supported hydraulically so that there is no distortion of its curved shape.

Casting optical glasses for telescopes is not a new business for Corning. The company made the mirror for the famous Hale telescope on Mount Palomar in California, the two-hundred-inch instrument that took nearly twenty years to build before its dedication in 1948, and, later, NASA's Hubble space telescope, which was stricken with mechanical and electronic problems, though not because of Corning's work.

Early in 1997, space-walking astronauts visited the telescope to make repairs and install two new instruments more advanced than anything yet used to track the life and death of stars. By May of the same year, the Hubble was performing in a way that more than atoned for all its past failures. It was with the use of Hubble before repairs that a team of scientists found proof of the existence of black holes in space, but with the two new instruments, a space telescope imaging spectograph and a near-infrared camera and multiobject spectrometer (STIS and NICMOS, respectively, in their acronymic lives), direct evidence of a black hole so large that its mass is three hundred million times that of the sun was recorded. Hubble also revealed the violence of an exploding star, and of a ring of gas swirling around the black hole at speeds up to 880,000 miles per hour. That all took place in the center of M84, a galaxy fifty million light years away from earth.

Corning's telescope mirrors were once made of a glass-ceramic, not unlike that of Pyrex cookware, but now the material used is a true glass labeled Ultra Low Expansion, or ULE, and given the code number 7971. To make it, silicon tetrachloride and titanium tetrachloride are treated in a flame hydrolysis process to form first a vapor and then particles of molten titanium silicate. Amassing over a period of one week, day and night, the particles fuse into a disk, or boule, of ULE glass the size of a tractor tire. With this glass, there are none of the contaminants of sand to be concerned with, and, equally important, the total thermal expansion of the glass is near zero. If the glass of a telescope mirror responds to heat, there are likely to be changes in the curvature of the glass, and that, of course, will result in distortion of the images. And at night, when the observatory dome is opened and the cold air of the mountain's summit rushes in, the glass must not recoil from it.

The world has had a love affair with telescopes since that day in 1609 when Galileo first demonstrated the instrument. In contrast to the twenty-seven-foot mirror blanks being cast now, Galileo's telescope was a mere two inches in diameter. But that was enough to bring detail to the fuzzy wash of the Milky Way and to peel back the cover of darkness from countless numbers of stars and lunar craters never before seen. As telescopes got larger, they could handle more light and bring up more of the unknown to view. Smallness is seldom a virtue in astronomy, but as mirrors and lenses grow larger they will start to distort because of their weight; because of the increase in lens size over a period of three centuries, lenses were replaced with concave mirrors as the heart of telescopes. It was a move from refracting to reflecting.

But with mirrors, too, size and weight constitute a threat to reliable operation of the telescope. Weight aside, the size of the glass determines how quickly it can heat or cool itself to the same temperature as the night air; the larger the glass, the more chance that its size may be altered during the heating or cooling to an extent where it can no longer focus light properly. Up until the early 1990s, the mirror of the largest optical telescope in the world measured six meters, or almost twenty feet, in diameter, and weighed forty-two tons. It was a Soviet installation, close to Zelenchukskaya, Russia, in the Caucasus Mountains, and its performance has been disappointing. Unlike the new Schott and Corning mirrors, the Soviet model is too bulky to respond quickly enough to changes in temperature, and that distorts the light as it is collected.

In the end, this has become a challenge to engineers and glassmakers, and they have started to respond. A promising possibility would have computers control the adjustment of large, very thin mirrors to temperature changes and other forces such as winds. Taking the light from a number of small telescopes and combining it for use is another strong possibility, now that the computer is available to coordinate the alignment of the instruments.

If incoming light is distorted, adjustments should be made to the mirror's surface to compensate. The changes must be made with millisecond speed, and one way to do that would be through the use of adaptive optics. The mirror would consist of many parts, and rather than respond to the incoming light as a single unit, the parts would be independently positioned by a computer.

One of the most radical departures from conventional telescope-mirror making thus far came in 1990, when the

Keck telescope was first turned to the skies. Named for W. M. Keck, an oilman from California whose foundation donated $70 million to the project, the telescope is located on Mauna Kea, Hawaii (that dead volcano may never spill lava again, but it is rumbling with astronomical activity), and it is an extraordinarily complicated instrument. Although the mirror is a staggering ten meters, or almost thirty-three feet, in diameter, it is segmented into thirty-six separate pieces, each of which is only three inches thick, resulting in a substantial weight reduction. The total weight of the Keck segmented mirror is 14.4 tons, as compared to 20 tons for the smaller mirror of the Hale telescope.

The three dozen segments of the Keck scope act as one in focusing starlight. Just making the small pieces was a major achievement, in that each one had to be polished to its own curvature, different from all the others. It all called for such precision that with some polishings, the glass had to be removed one molecule at a time by bombarding it with argon ions; and in aligning the segments, there was no more than a millionth of an inch to play with. The mirror rests on a cushion of oil, and to maintain proper curvature, electronic sensors send data regarding the position of each segment to a computerized system, and that in turn orders hydraulic levers to move the individual pieces as needed. The computer's signals are sent twice every second. Small wonder that a number of scientists expressed doubt that such an instrument could function without a procession of problems. But it has been a success, pointing the way for telescope mirrors of even larger size and still of acceptable weight.

Dr. James Angel, a professor of astronomy at the University of Arizona, in Tucson, conceived the idea of making a mirror for a ground-based telescope by having the glass cast as one unit, but infused with air bubbles for lightness. A pot was constructed, and the molten glass was poured in to take shape around the blocks that made a form for the air bubbles. The blocks were raised so that the molten glass could get under them to form a solid underside for the mirror. It was done with a spinning process, and when it was over, Dr. Angel and his colleagues had made the largest telescope mirror of monolithic cast in the world. "It is twenty-seven feet in diameter, three feet thick, and weighs just sixteen tons," Dr. Angel said. "It is mostly air, but the honeycombed glass is quite rigid."

This mirror was made for a telescope that will resemble a gigantic pair of binoculars; the instrument, named (no surprise) the Large Binocular Telescope, will be positioned on a site in Arizona. The twin mirror will also be cast by Dr. Angel and his coworkers, in the same pot at the same place—under the football stadium at the university.

By taking a mirror that is both lightweight and thick, Dr. Angel met a goal that has long eluded astronomers and glassmakers. "In a way, we built this mirror on the same principles that the great medieval cathedrals were built," Dr. Angel said. "They were designed to withstand gravity and not collapse under their own weight, and they had a limit of a hundred and forty feet in height so that they didn't collapse in strong winds. When the wind blows, this mirror will keep its shape. Once you get the glass right, it has this amazing ability to resist age. Look at the Mount Wilson Observatory telescope; it's a hundred years old, and the glass still has the same shape and polish."

Unsurprisingly, Dr. Angel is the recipient of a $330,000 MacArthur Foundation "genius" award, granted in part because of his work in developing the new method of casting glass for telescope mirrors.

However the mirrors are made, it seems certain that if there comes a time when the heavens give up a secret of another place like ours—an alien world where extraterrestrial life works and sleeps and has trouble getting a plumber—it will be optical glass that delivers the news; the same material, but with refinement, that was used to make an unguent jar in the thirteenth century B.C. And will it be said in the year 6000 that glass has spanned even another four thousand years, this time to become sequined with the light of other suns?

It is always a struggle with optical glass to stay thin. It has to do with the refractive index of the material, a calculation whereby the speed of light in air is compared to the speed of it in the glass. So when light strikes glass, the amount of deflection from the path on which it enters will be determined by the refractive index of the glass: With a high refractive index, the more the light is deflected. Glass of a low refractive index must be curved on the surface to deflect the light, and that means the optic must be made of thicker glass. With lenses and prisms of optical image systems, the light is broken at the surface of the glass, and for that there are glasses that spread the light wide, and glasses that break white light into its spectrum of colors.

The optical glass most familiar to people the world over is, of course, the lens for spectacles. To judge the importance of what the eyesight-restoring glass has done for humanity, give thought to what it must have been like before the invention of spectacles: men and women—children, even—with no escape from a hazy world, forever frowning under the strain of trying to focus; nearsighted readers with the scrolls touching their noses; astigmatic charioteers running into each other.

The origins of eyeglasses are not at all clear. There are Italians who make a strong case for Venice and the influence of the *cristallieri*, the craftsmen who first worked with rock crystal, or quartz. In the thirteenth century, a *capitolare*, or collection of ordinances governing the trade, was drawn up by a panel of judges, and it was stipulated that clear uncolored glass was not to be passed off as the more expensive quartz. Specifically mentioned in that regard were *lapides ad legendum,* or, more or less, "stones for reading," meaning a magnifying lens. Also cited in that ordinance are *roidi da ogli*—"spectacle lenses"—and in a fourteenth-century update of the ordinances, in which Latin gives way to Italian, there is a reference to, in translation, "making rounds of glass for reading spectacles."

Roger Bacon, the English philosopher and scientist, as well as Franciscan monk, wrote about spectacles during his very productive lifetime (1220–1292), but of course the poor man's foresight was interpreted as black magic. As for eyeglasses in America, they arrived in the New World, so it has been recorded by the American Optometric Association, in 1620, on the nose of Mayflower passenger Peter Brown. More than two hundred years later, Theodore Roosevelt packed twelve pairs of pince-nez to take off to the Spanish-American War. The history of spectacles is blinding with such flash cards of minutiae.

By the time of that August in 1609 when Galileo was atop the bell tower in Venice's San Marco square, demonstrating his telescope, the eyeglass business was in full bloom, although there was little of today's cosmetic considerations given to use of the vision aid. If someone wanted to buy a pair of glasses, he or she usually went to the local haberdasher's shop and searched among the notions.

For most wearers of eyeglasses now, it isn't enough just to have faulty vision corrected. The lenses must be as light and thin as possible; they must be scratch-resistant. And, most of all, the framed glasses must be stylish and look good on the wearer. It is a lot to ask for, but not out of line with what the seller is asking for in return. As many as 145 million Americans wear eyeglasses or contact lenses, and so dispensing eyewear is a vibrant business in which sales of several hundred dollars and more each are cranked out over the counter. About 70 percent of the prescriptions can be filled with ground lenses already in stock. Up until the early 1970s, it was glass that carried much of that trade—but the optical glass of the lenses was often thick and heavy, and always with no immunity whatsoever against a rock propelled eyeward by a lawn mower.

In 1972, the Food and Drug Administration decreed that lenses made of glass must be a minimum of 2.2 millimeters (.0866 inches) thick, and that, in effect, sounded the death knell for the use of optical glass in spectacles after being the material of choice for more than five centuries. By 1996, plastic lenses were going into 71 percent of the prescription eyeglasses being sold in the United States.

"Before the FDA ruling, a higher nearsighted correction could be made with a lens of point-five millimeter thickness, but then, overnight, it had to be two-point-two," said Joe Bruneni, a consultant to the Optical Laboratories Association. "The weight of that correcting lens increased four or five times." Not unlike the way a shopper prefers food and drinks in lighter containers, the eyeglass buyer places a value on less weight and thickness. And that is not in the best interest of glass. The FDA, however, acted as it did not because of concern for comfort, but because of worries about safety. Eyes can be seriously injured if a glass lens shatters; that had happened often enough for lobbyists representing associations of blind people and others to bring pressure for government intervention.

Outside of the United States—in Europe, especially—the use of glass in spectacles remains high. One of the attractions of glass as a lens has always been its resistance to scratching, but even that has become less important with the introduction of plastics that can remain scratch-free over long periods of time. Plastic was first used in eyeglasses in the 1930s, though it was of low quality. By the late 1940s, the plastic had improved, but a lens made of it could still be twilled with scratches. It took thirty years before it became refined enough to pose a serious challenge to the long dominance of glass. And now lenses are being made of a polycarbonate that can protect the eyes from a line-drive shot off a baseball bat.

Glass lenses retain the advantage over plastic of offering better protection against infrared radiation. They also put up more resistance to chemicals, and, of course, no matter what coatings are applied to plastic, glass will continue to be less prone to scratching. Glass or plastic is but one of many choices to be made by the eyewear customer today. There are the lenses: polarizing, high refractive index (allowing for thinness of lens), photochromic, or even aspheric, meant to soften the startling goitered look of eyeballs behind lenses of high magnification. There are single-vision lenses, bifocal lenses, trifocal lenses, and one called "progressive addition lenses," whereby there are focals enough to handle a wide range of eyesight problems. With that decided, then this: spectacles or contact lenses?

Almost all contact lenses are made of plastic, as they were from the start, just after World War II. The Plexiglas used to make canopies for the cockpits of fighter planes was found—a find after the tragic fact of casualties in war—to remain fairly inert when in the human body, and so contact lenses were made of that plastic for a time. Today, there are disposable contact lenses, and ones of different colors, but what has not evolved is a contact lens with a tracking device to pinpoint its lie in a thick rug.

Plastic has also intruded by now on the glass science of photochromism, whereby the lens changes color to compensate for the amount of light available. The lightweight plastic eyewear that darkens in sunlight has been hugely successful in the marketplace, and there can be no question that the trend is disheartening for glass people. But at least one giant among them long ago put a finger to the wind: The leading firm in the production and marketing of photochromic opticals in plastic is called Transitions Optical, Inc., the majority of which is owned by PPG Industries, the glassmaking concern.

If glass eyewear is all but finished in the United States, it has strengthened its hold worldwide on high-tech optics. Most plastics in heavy use today cannot stand up to high temperatures, and that is a crippling deficiency when it comes to being molded or otherwise shaped. Here, for example, is the usual procedure for machining a high-precision optical lens, as followed at Schott's: The glass is melted, dropped into the mold, pressed, and sent to the annealing oven for cooling down. It is then milled on both sides, cleaned, laminated, ground and polished on one side, cleaned and laminated again, ground and polished on the other side, centered, and cleaned again. Most plastic would have to drop out at the melt.

Not all lenses are individually ground and polished. There is a heavy demand now for small precision lenses, and to grind and polish such tiny lenses in the usual method is exceedingly difficult. For one thing, the lenses are aspheric in shape (a slight deviation from the spherical shape which can cause problems with focusing), and grinding may alter the shape; also, mass production of the small lenses would not be possible. To solve the problem, glassmakers, including Corning and Schott, turned to a process of compression. To fabricate quickly, but with no sacrifice of quality, a glass preform is heated and pressed between two molds representing both sides of the lens. The molds have been ground to within precise tolerances for accuracy. When released from the assembly, the glass has taken on the same contours and surface profile of the molds. It is not so simple as that, of course, since temperature and timing of the preform's contacts with the molds are factors with no room for error. Also, materials of different thermal reactions are used in the process, and the expansions of those materials under heat must be aligned for the best possible imprint of the mold.

The wide-ranging need for high-precision microlenses was kicked off by development of the digital camera in the early 1980s. Now such lenses are used in camcorders, audio compact disks, and videodisks. There are molded lenses in the bar code scanners at the supermarket checkout stands, and, with an application of more style and substance than decoding the price of a can of tomato soup, in the couplers that link lasers to diodes and then to fiber optics. Some other systems are for optical storage systems and printers.

Once again, then, glass has moved in lockstep with the advancement of science, this time as a small, molded aspheric lens of high precision.

The plastics industry is attempting to make inroads in the molded lens business, but so far plastic hasn't proven itself to be a completely suitable material for that. Plastics are soft, and when the temperature rises, there can be problems with their refractive index. A precision lens, as nabobs of the plastic business must know, is not a container for soda pop.

For all of that, the great strength of optical glass continues to rest in its ability to raise the invisible to light. A new generation of optical instruments is emerging that can provide detailed imaging of the inner workings of cells. Called near-field scanning optical microscopes, they can harness what one scientist calls "the power of photons" to resolve images down to approximately one two-millionths of an inch. As instruments of masterly technology, they have revoked the principle of physics that said resolution of any-thing less than half the wavelength of visible light was not possible.

For an even stronger performance, there is the scanning sensor microscope developed by Schott for the testing of surface structures. Using laser light, it has the ability to image individual atoms. A sample to be studied by this in-strument is scanned by having a sharp-tipped sensor moved across its surface at a distance of just a few nanometers. The tip itself is a glass fiber drawn extremely thin and coated with aluminum. The light of a laser is directed up from below, going first through a glass optical lens to be focused, and then through the study sample and into the fiber-optic sensor. As the sensor scans, the light is constructing an image which can be a picture of the interior of a cell.

Resolved down to just two centimeters with the scanning sensor microscope, the crystal structure of a grain of salt appears as a dead ringer for a piece of Berber carpet as seen by the naked eye. And it can resolve down even more to reveal the 0.4-nanometer spacing of the atoms.

As with telescope technology, the latter part of the twenti-eth century has been a period of great revival of microscope technology, with near-field scanning the featured attraction. Other than showing us what we haven't seen before, new optical instruments have capabilities such as reading and writing astonishing amounts of data on a computer disk. Scientists talk, too, of using the near-field fiber-optic probe to write circuitry on silicon chips.

In more of a contact sport for optical glass, it is being used to test the energy of certain subatomic particles, such as electrons; scientists put them through optical glass laced with lead (to absorb radiation). In another application, the lead-containing glass is used in the four-foot-thick windows of nuclear power facilities. For the welder who works in a burst of sparks, there is optical glass for the window of the eyeshield. There are convex rearview mirrors for automobiles, and high-precision lenses for cameras.

In the making of optical glasses, there are wide selections to be made in regard to ingredients for the melt, but there can be no deviation from the essential need of maintaining low tolerances in the glasses. Without precision, the glass is but a glass, but when ground and polished to a smoothness where even an atom dare not rise unremarked, then it becomes like a talisman on which is borne some of the truth of science.

And nothing so becomes it as light.

Chapter Ten

Nightmares Buried in Glass

As glass can reach out to the stars, it can also carry our nightmares deep into the earth.

In the piney woods of South Carolina, technicians are making glass with radioactive waste. Jacketed in steel, it will be hidden away forever—the safest and most acceptable answer yet to the dilemma of what to do with that dreaded refuse.

Because glass retains its physical and chemical characteristics when fused with other ingredients in the melt, a block of the hazardous nuclear-waste glass can be encapsulated in a stainless-steel canister and buried in the ground without fear of leakage. The United States was late in coming to this method of waste disposal, as it had already been put to use in Canada, France, Belgium, and the United Kingdom.

In the 1950s, a time of rising tensions between the United States and what was then the Soviet Union, work began on developing a 310-square-mile chunk of South Carolina farmland near where the Savannah River marks the boundary with Georgia, close to the city of Augusta where the Masters is held each year, attracting the best golfers in the world to a course of famed beauty and challenge. It is an area still rich with the culture of a genteel South—of breathless ladies cutting flowers for the many rooms of large white houses, and thrice-married men named Clay slightly tipsy in the clubhouse bar.

A great tide of equipment and people rolled in, mandated to build what was needed to have this place bristling with the making of nuclear weapons. The production would be of tritium (a form of hydrogen) and plutonium 239. Some of the plutonium would be earmarked for medical use and for the space program. Five nuclear reactors were built, along with two chemical separation plants, known as the "canyons," and a facility for extracting heavy water. The riverfront site was secured, sealed off from the world, ringed with armed guards more stern-faced than Oliver Cromwell. The work went on there for close to thirty years—irradiating, separating the usable from the waste, refining, shipping. From 1953 to 1988, the production of plutonium at the site totaled something like thirty-six metric tons. And all the while, much of the waste was piling up, not only in Savannah, but in other weapons-making sites across the country— piling up and piling up until now the government estimates it will take at least thirty years to clean it all up, at a cost of $100 billion.

It wasn't until 1989 that the government started to pull back the cloak of secrecy from the Savannah River site and what was there: thirty-four million gallons of radioactive waste.

The low-level wastes are in the form of a salt solution similar to fertilizer. This material is being solidified as cement and then placed in vaults that go in the ground under a clay cap. Throughout the history of the Savannah River site, there have been ongoing disposal programs, such as the burial of radioactive solvents in underground tanks. One

burial site for wastes covers seventy-six acres, and twenty
years passed before it was full. (It was capped for good
in 1996, a procedure involving what was termed "dynamic
compaction of the materials." In a somewhat startling demonstration of Flinstonian technology, a crane was brought in
to lift a twenty-ton weight thirty-three feet in the air and
then release it to fall onto the waste site, over and over
again—dynamic compaction.)

All of that has been good enough for some low-level materials, but the 2.4 million gallons of the much more hazardous
waste that remain—the unspeakable, the deadly sludge from
the bottom of that dark pit of distrust and tension we called
the cold war, waste emitting radiation from cesium and
strontium isotopes—requires fail-safe disposal. There is one
method of doing that in which safety and permanency are
all but sworn in blood. It is called vitrification; it means that
the best way to dispose of the waste is to transform it into
glass. The decision to go that way was made only after a
great deal of research, and the findings were that the waste
as glass would be ten thousand times more durable than
waste as cement or any other form, thus reducing the risk
of leakage over the many, many years it will take for the
waste to become anything but deadly.

"Vitrification is not new," said a representative of the
Westinghouse Savannah River Company, operator and
manager of the facility for the U.S. Department of Energy.
"Research into the process was going on at MIT more than
forty years ago. The big attraction to turning the waste into
glass is that nothing leaches from glass. But understand,
we're not going to just encapsulate the waste in the glass.
The waste is the glass itself."

It will take at least fifteen years to process all the material
at the site. And, after two years of test runs and other start-
up procedures, the work has begun. In the first year of oper-

ation, starting in 1996, some seventy canisters, each ten feet tall and two feet in diameter, have been filled with glass—a black shiny glass, the deathly reflection of what it was.

The five reactors are all shut down now, although one remains on "cold standby," meaning it can be reactivated more quickly than the others. But that is not likely to happen. While the cold war has ended, the United States continues to include nuclear weapons in its arsenal for defense, and that means that the tritium in the warheads has to be replenished constantly, since the man-made gas has a half-life of just 12.33 years. There is enough of the material available now to last until well into the first decade of the twenty-first century, but after then new supplies will be needed. By that time, a new technology for making tritium will likely have been brought into use. Instead of a reactor, a linear accelerator will be used, one now being designed by a multilaboratory team, including scientists from the University of California's Lawrence Livermore National Laboratory. In the accelerator process, subatomic particles are made to travel at nearly the speed of light, so that when they hit a target, additional atomic particles are freed. And on and on it goes, reaction after reaction, until finally some of the neutrons are trapped in a blanket of helium, and it is then that the tritium takes shape. The accelerator will be about a mile long, underground.

So the five heavy-water reactors at the Savannah River site stand now in weedy solitude, away from the vibrant activity at the new building constructed for vitrification of the hot waste. It is more a fortress than just a building, in that sixty-nine thousand cubic yards of concrete and thirteen thousand tons of steel went into the making of it. "This building will resist a steel pipe traveling a hundred miles an hour," an engineer on the site said. The facility is fully automated, for this truly is a glass to see through not only darkly but at a distance. Even the air that leaves the building goes through an eight-foot-thick sand filter to make certain that contaminants do not escape. There is no waste processing facility in the world larger than this one, and, with a cost of more than one billion dollars, none so expensive.

The sludge waste is mixed with water and fine particles of a borosilicate glass that is 50 percent silica, 16 percent soda-lime, 24 percent metallic oxides, and 10 percent boron (a fluxing agent). Joule heat—electrical current passing through two electrodes—is used for the melting. The batch consists of 51 percent water, 25 percent glass, and 24 percent waste and is fed into the ceramic-lined melter at a rate of one gallon a minute. The waste itself has been mixed with a salt derived from benzene fluid. There is strong resistance by the radioactive material, now a slurry, to the current at first, but that fades once the melting starts at a temperature of 2,100°F. As the slurry is fed from the top, it forms a crust on the melt pool as the water is evaporated and removed. Melting occurs on the bottom of the crust, or "cold cap," to form the borosilicate glass.

It takes sixty hours for the complete melt, a time when essential parts of the most destructive force ever created by humans fall under the spell of simple glass.

From the melter, the mixture goes into the canister, there to harden after the sixteen hours it takes to fill the 3,700-pound capacity of the steel jacket. What remains is to decontaminate the exterior of the canister by blasting it with wet sand until one thousandth of an inch of the steel is removed. A hole in the top of the canister is too small to take the steel plug meant to go there, but when a force of seventy-five thousand pounds is applied to the plug by treating it for one and a half seconds with 230,000 amps of electrical current, the plug fits with a fastness to last for eternity, or at least for the eight thousand to ten thousand years it will take for the black glass to lose all radioactivity.

Workers stand behind thick walls, peering in at the operation through windows of glass more than four feet thick. A crane is being operated by a joystick from a distance. There are safety devices everywhere—bells that ring and lights that flash when something is amiss.

The men and women here dressed in yellow protective clothing are involved in radiological work. At one time the clothing was white, and when that changed, the old issues were laundered and shipped to workers cleaning up the disaster-stricken Chernobyl nuclear power plant in Russia. In keeping with the post–cold war policy of openness at Savannah, it was disclosed, before anyone could ask, that the used protective clothing was sent with some contamination still in it, but at levels "low enough to allow re-use by radiological workers and . . . within radiological release guidelines."

On May 17, 1996, the filling of the first canister with radioactive glass at Savannah was completed. Thomas P. Grumbly, undersecretary of energy in the Clinton administration, was there to say, "I consider this facility to be [the Department of Energy's] most important environmental cleanup project, not only because it stabilizes Savannah River site waste, but because of the potential application of this state-of-the-art technology at other sites." On hand to move the five-thousand-pound canister to the storage building was a behemoth that stood eighteen feet high and weighed 235,000 pounds, the "shielded canister transporter," made for this work only. It carries one canister at a time to a concrete and steel vault placed deep in the rich soil where once plowshares carved furrows for corn and beans.

The vault can hold 2,300 canisters of the black glass. The government wants to send those and all the others to follow out to Nevada, there to be buried in 6,545-foot-high Yucca Mountain, on a site where nuclear weapons are tested. However, that plan remains under study (and under the hammering objections of many residents of Nevada).

"All of this was built down here because the Russians exploded a hydrogen bomb," Ned Bider, a chemist who came to Savannah in 1965, said. "We built our bombs, and that's over, so the whole idea now is to dissolve a dangerous waste and then put it in a safer form. We continue to look at different ways to do that, and vitrification—glass—continues to be one of the best."

In studying possible methods of dealing with high-level wastes, scientists came down strongly on the side of glass mainly because of the material's durability. Dr. Carol Jantzen, a scientist at Savannah, is responsible for much of the pioneering work in getting vitrification accepted as the preferred method of hazardous waste disposal in the United States. Recognizing the intrusion of groundwater can set free radioactive nuclides buried in the ground, Dr. Jantzen said, "It is important that nuclear waste glasses be stable in the presence of groundwaters for very long periods of time. We only have to look at the natural glasses, such as obsidians. They are millions of years old. There are synthetic glasses like those of medieval windows, and they are very old. That should demonstrate the potential long-range performance of nuclear-waste glass."

As forbidding as the vitrification work at the Savannah River site is, that place seems to be infused with a sense of renewal. They are finally burying the bad waste of what they made for many years, and it is like burying the past, too. Some took it as a good sign that two yellow-bellied slider turtles were found on the site in 1996; one still bore the markings put on it when it was captured and released there twenty-eight years earlier. Indeed, the site grounds have always been an important wildlife habitat—seven species on the endangered list live there.

Canada began vitrifying nuclear waste in 1957, England in 1962. In Australia, the commonwealth government sought ways to dispose of radioactive wastes buried at a nuclear test range in the south of the country. Excavation and vault burial was considered, and so was chemical stabilization. But when vitrification tests were conducted, the results made it clear that anything less than that would pose added hazard in dealing with the material. With vitrification, the tests showed, 99.9999 percent of the plutonium present in the material would be retained in the melt. The government awarded a $21 million contract for vitrification. In France, the Commissariat à l'Energie Atomique (CEA) was working to achieve a satisfactory method of disposing of high-level waste, a problem of deep concern in a country where the development of nuclear energy has been intense. The concern was for the wastes from power plants, but, to bring light to a house or destruction to an enemy, the basic process is all more or less the same: the nuclear chain reaction, the conversion of uranium into plutonium and other elements, and the buildup of spent fuels containing radioactive fission products. With some of these radionuclides, the problem of disposal is short-lived. Iodine 131, for example, has a half-life of just eight days. But there's also iodine 129, with a half-life of sixteen million years. As much as 99 percent of radioactive waste is short-lived, but, alas, that 99 percent represents less than 1 percent of the radioactivity itself.

At first, the CEA experimented with transforming the wastes into synthetic mica crystals. That didn't work, so the emphasis switched to vitrification and glass, and that's the way the wastes have been handled there since the 1960s. A cylinder of vitrified waste from a power plant, measuring no more than an inch and a half tall and an inch in diameter, contains the radioactive remains of the atomic energy it would take to provide a lifetime of electricity for a family of four.

In Europe, too, the canisters await final disposal deep in the earth, but there have been delays, since nothing so raises public apprehension as the thought of radioactive material in the ground. Meanwhile, the inclination to look to outer space or the sea for answers to the problems of hunger and health (as well as burdensome populations and poor television reception) has brought forth proposals for launching nuclear wastes into space, and for designing the steel jackets that encase the vitrified material with torpedolike fins so that they can be propelled into holes drilled in the ocean floor.

It is glass, though, that is slowly clearing out the stores of high-level hazardous material in the world. The tonnage of black glass in steel canisters is increasing every day. And it could all be just a prelude to what is to come as a result of agreement by Russia and the United States to eliminate large stores of nuclear weapons from their arsenals. In addition to the wastes that piled up during the making of the weapons, the uranium and fifty metric tons of plutonium in the warheads themselves and in reserve will have to be disposed of, and done so quickly to guard against the material getting into the possession of terrorist organizations. If vitrification is the method chosen to dispose of the warhead material—that seems certain, since the Environmental Protection Agency has declared vitrification to be the "best demonstrated available technology" for treating heavy metals and high-level radioactive wastes—then it will be complete, with both the waste and waste-maker made molten as glass, and poured into the steel cylinder as if in a decanting of reborn trust among major powers.

Sometimes it seems you can hear the old secrets of Savannah popping out from behind closed and guarded doors as you walk on those grounds. It is now revealed, for example, that a certain solution was stored for many years in a building called the F Canyon. On the low-to-high-level scale of radioactivity, this brew reaches above that, for it contains isotopes of the elements americium and curium, both products of plutonium being bombarded with high-energy neutrons. Ironically, it was not produced for weaponry, but for medical and other peaceful purposes.

The material was needed at the Department of Energy's Oak Ridge facility, in Tennessee, for there such radionuclides are made available for research. But the federal government will not allow radioactive materials in liquid form to be shipped. They decided then to vitrify the twenty-eight pounds of the solution and send it as glass. Because of the intense levels of radiation, it was determined that the work could only be done by lowering a melter system on a frame into a special area of the F Canyon building and routing the solution to it. Working at well-shielded controls, at a distance, operators mixed the material with glass, conducted the melt, and filled the first of forty small steel canisters. After arriving in Oak Ridge, the glass is to be treated with nitric acid in order to recover the americium-curium content.

Because of the ban on shipment of liquid hazardous wastes, the state-of-the-art vitrification facility at the Savannah River site processes only the material stored there. What, then, of the contaminated stocks at Oak Ridge and other non-Savannah sites?

The answer may be observed now and then on the nation's highways, as ten tractor-trailer rigs haul the modules that make up the Transportable Vitrification System. It is the first of its kind (although, of course, as a portable waste eliminator, the Porta-John was at the construction site long before the TVS was even an idea).

Now at Oak Ridge, the system has the capacity to treat a

little more than a million pounds of low-level waste at a rate of four to six tons an hour. More such portable units are likely to be built now, to be moved from here to there, calling on places where there are thick walls imprinted with the cautionary logos of the atom-smashing business. Operating these portable units is a business on the rise, and there are young companies striving to hold on as they look ahead to a time when other hazardous materials will be disposed of as glass.

And beyond that, they see the commercial possibilities for the waste glass. The glass made of nuclear products remains hazardous after vitrification, but that isn't true for, say, glass made of asbestos.

Asbestos causes cancer, as many as ten thousand cases a year in the United States, and few of the victims survive. Unlike nuclear wastes, asbestos is no longer hazardous after it has been vitrified, since the tiny fibers that could be inhaled and set off the cancer break down into harmless oxide compounds during the vitrification melt process. Being harmless, the glass need not be buried deep in a mountain, but could go into a landfill. The usual procedure for getting rid of old asbestos is to put it in a double-sealed bag and put it in a landfill. As glass, the asbestos would take up much less room, but since it is an issue of space rather than safety, cost factors, in which vitrification cannot compete, are all-important. The cost of glassifying high-level nuclear wastes (about $750 a pound) was less of a factor in the decision to use that process because of the overriding concern for safety.

Duratek and other companies have looked ahead to the possibilities of finding markets for such waste-containing glass (though not, of course, the hot glass of high-level nuclear wastes). If the glass that was cancer-causing asbestos is now benign, why not make flower vases of it—why, indeed, not have casts of the black glass meant to hold some ice-pink camellias, or a tuft of yellow coreopsis, in celebration of the cleansing rebirth of waste? That's a possibility, as is using the glass for road construction—as an alternative to gravel, and as the "glasphalt" substitute for asphalt, but that is likely to take a long time since there is an abundance of ordinary glass scrap, or cullet, available for that growing use.

There are many uses for the nonhazardous vitrified waste, but the route to riches for businesspeople leads through what can seem like a minefield. Much depends on the environmental policies of federal and local governing bodies. As the average shopper in the United States is reluctant to buy organically grown fruits and vegetables because they do not have the good color and smoothness of those produced in the heavily treated field, so too are drivers uneasy about having the tires of their cars roll over roads that sparkle with glass. Glass-ceramic is a material of golden promise, but to make that of the vitrified waste requires a wealth of engineering and scientific procedures; it is not something to be done in the shop of a small company struggling to meet its debts. Also, it will require masterful salesmanship to sell vitrified waste as a substitute for other materials already in place.

Vitrification is a process with a wide reach. Other than high-level nuclear wastes and asbestos, there is the ash discard from municipal incinerators to be glassified, and the industrial legacy of wastes that wash into rivers and streams. I am thinking here of the petrochemical industry in Louisiana, along the 150-mile-long Mississippi River corridor between New Orleans and Baton Rouge. Of the one billion pounds of weed and bug killers used throughout the United States each year, many are made in plants along the river, as are fertilizers and petroleum products. A by-product of fertilizer production is a waste called gypsum, but not the gypsum used as wallboard. This gypsum contains sulfuric and phosphoric acids, and even radium. The practice has been to let the gypsum pile up, sometimes in a heap sixty feet high and more than a mile square. Rain has washed it off, to be carried into the Mississippi, the same river from which New Orleans gets its drinking water. And so here is a case for the future of vitrification: Glassify the gypsum, and let it shine in that place where sunlight is blocked by the putrid emissions of the stacks that rise along the banks of the nation's mother stream. The ash from the burning of massive loads of medical wastes is another eligible partner for that molten embrace when there is an everlasting bonding with glass. "We can take a large box of those ashes and turn them into a piece of glass no larger than a plum," a scientist at the Savannah River site said.

Radioactive material has been found in the sludge at the bottom of New York Harbor, and it is believed that vitrification can correct that problem. Old lead paint scraped from walls holds metallic poisons. Vitrify it. But of all the targets other than high-level nuclear wastes, none is raised to such clear sight as the industrial legacy of contaminated soil. There are even references now to the "contaminated soils market," estimated to have a worth of between $200 million and $300 million during the next twenty-five years.

Superfund, a creation of the federal government, is a resource with which to clean up hazardous-waste sites in the United States that have been abandoned or otherwise need emergency controls. By its formal name, it is the Comprehensive Environmental Response, Compensation and Liability Act of 1980, and is administered by the Environmental Protection Agency (EPA). Under authority of the act, the agency moves in when those responsible for the hazardous situation fail to take remedial action and simply skip out, often leaving behind many acres of soil contaminated with arsenic, mercury, and dioxin. Or the site may be in a farm area, where the pesticide-laced soil can set a cancerous tumor to growing about as easily as it can a squash.

The bad soil can be incinerated, but that does not deal with the metallic contaminants. With incineration, the combustion of fuels needed to obtain the necessary temperatures itself leaves a large volume of by-products to dispose of as well; combustion is not necessary for the vitrification process. Incineration and vitrification aside, there are other things that can be done with contaminated soil, but nothing so final and complete as turning it into glass.

Realizing the usefulness of vitrification in dealing with hazardous materials, the Department of Energy, in conjunction with a private company called Battelle Pacific Northwest Laboratories, of Richland, Washington, developed a technology for vitrification of contaminated soil *in situ*. The technology was made openly available but attracted no interest. Deciding to produce as well as invent, Battelle formed a satellite company called Geosafe and went into the business of turning contaminated soil and sludge into glass.

How it works: Four electrodes are set in the ground to mark a section for treatment, one something like thirty feet in diameter and eighteen feet deep. Electric current is introduced into a starter path of graphite and glass particles laid down on the surface. The heat will reach temperatures over 3,600°F, and the molten mass of the soil itself becomes the primary conductor of the electricity, thus allowing the process to continue. During that time, a 20 to 40 percent reduction in the void volume or air spaces of the soil occurs. Gases escaping from the soil are trapped by a collection hood and diverted to a treatment unit for cooling and cleaning before being released. In the glowing ground, meanwhile, the heat is removing and paralyzing the organic compounds of the harmful chemicals, while any contaminating metals present are being wrapped into the forming glass as oxides.

Other than not having to move the contaminated soil, *in situ* vitrification offers the benefits of processing organic, inorganic, and radioactive contaminants simultaneously and providing high tolerance for debris, such as metals and concrete.

Vitrification by a mobile unit at a Superfund site was first used at Grand Ledge, Michigan, in 1990. From 1945 to 1979, a company there called Parsons Chemical Works produced pesticides, and during those thirty-four years, a hellish stew was mixed in the ground: mercury, lead, DDT, dioxin, even some arsenic. The contamination involved three sites, covering some three thousand cubic yards of soil in all. A $1.7 million contract to clean up the site was awarded by the federal government to Geosafe.

Geosafe conducted eight individual melts at the Parsons sites, and then the glass of all those melts was melted again, this time together, to form a glass monolith with ten times the strength of plain concrete. It is 13 feet deep, 160 feet long, and 70 to 90 feet wide, resting under three feet of backfill. Of course, the glass can be removed, and the best way to do it is to mount a hydraulic hammer on the arm of a backhoe and chip away. No dust escapes. But if left in the ground, it will likely survive as long as the material it most resembles, both physically and chemically, obsidian, believed to have an average life span of eighteen million years.

"The technology has worked very well," said Matthew Haass, a Geosafe official. "We've made money on the three Superfund projects we've worked on, and that part is encouraging. But in general, the environmental market is not good for vendors at this time." That was in 1997, when the Superfund program faced hard questioning concerning money spent and gains made. *In situ* vitrification of contaminated soil costs about $425 a ton, and that's more than it costs, but not much more, to dispose of it by conventional methods, such as incineration. But it's not an easy sell to go with the best if it costs more—not when public funds are involved, and certainly not when chemical and petroleum companies are made to clean up their own mess.

Hardly had the vitrification work begun at Savannah before tests were under way to gauge the potential of a new process to make glass of wastes. The technology is not new—in a way, it is as old as lightning. The tool is a plasma torch, a magical wand of heat more intense than that on the surface of the sun.

In this technology, a powerful electric current is applied to break down a rarefied, semineutral gas, working on the electrons and atomic nuclei and other components not with combustion, but with a spark, to make the gas—now a plasma—hot beyond easy comprehension. The process has many similarities to arc welding and to the electric arc furnace, in use since 1897.

The plasma torch is capable of reaching temperatures in excess of 18,000°F. With that heat, it can glassify contaminated soil quickly and with authority, but at a price few are willing to pay. It can do nonhazardous wastes more cheaply, but at about $100 a ton, the method is still too dear to compete with landfills. In Japan, however, there are two commercial plants making glass out of the ashes of incinerated municipal wastes using the plasma process. The glass is sold for use as paving stones and bricks, and in construction. And in one plant in France, a hundred tons of asbestos are being laid to harmless, glassy rest every week using plasma energy.

"As soon as it is legislated that disposal of hazardous materials in landfills will no longer be allowed, then we will become competitive in the United States," said an official of the Plasma Energy Corporation of Raleigh, North Carolina, one of the three or four companies in the world who make the plasma torch systems. Currently the major use of the torch is for independent control of the temperature needed for the processing of steel.

At the Savannah test demonstration of the plasma torch, the instrument was lowered seven feet into the ground, to the bottom of a hole ten inches in diameter. When activated and raised slowly to the surface, the flame left a plug of glass in its wake that measured four feet in diameter. In a way, it was like lightning striking a beach to leave behind an offering of fulgurites.

With the economic barrier breached, the use of plasma technology for waste removal through vitrification should gain more favor. Rather than wait for the ash, the process may be applied to bulk waste as it is received at the processing center, making glass of it there. Looking beyond that, even, who can say that a plasma torch will not one day become a common household appliance: a sprinkling of glass, a push of button, and good-bye chicken bones and cottage cheese gone sour?

Vitrification was never an option for treating the radiation-hot sands of Bikini Atoll, but I don't know why. Forty years and twenty-three nuclear explosions after the gentle Bikinians were evacuated from their small island, a group of elders returned to hear a scientist from the United States explain what could be done to expel the cesium 137 from the sand. The radioactive element was the legacy of the first test of a deliverable hydrogen bomb. Called "Bravo," it created an explosion in the lagoon equivalent to 15 million tons of TNT—enough of the explosive to fill a freight train reaching from one coast of the North American continent to the other. It is the most powerful weapon ever activated by the United States. Tragically, at the time of the shot, the winds shifted and Bikini and other islands received radioactive fallout.

The island was not so hazardous in 1986 that a person couldn't visit there, going ashore on a surf approach and drying quickly in the harsh sunlight. The elders moved around the islands, raising memories from an old carving in a palm tree, a cemetery headstone in the grip of a ropy vine. They were told that the top twelve inches of the soil on the island had to be removed or treated, most likely by applying fertilizer rich in potassium to the ground (in time, a decision is likely to be made to use the fertilizer). Potassium and cesium contain similar molecules, and when the fertilizer is put on the soil, the plants dine on that—in effect fooled into thinking the intake is still that of cesium. The cesium remains in the soil, but blocked from entering the banana trees and other eatable growths, it is not of significant danger.

Since that day of the elders' return to the island, I have built on a vision: The twelve inches of hot soil was vitrified, and wounded Bikini became an island of glass, a monolith risen from the sea. And it will remain in place forever, sleek and glinting, and so in harmony with the jewel-like days out there.

The rate of use of vitrification in the United States is being paced by the diminution of landfill space. As long as the ground will hold the waste and it is cheaper to dispose of our refuse that way, most nonhazardous wastes will be buried; much of the rest will be incinerated. So for now, glassifying waste in this country is more or less restricted to materials with fairly high levels of contamination.

Glass is a remarkable material in that, for one thing, it can swing between serving as something as sinister as a black embodiment of nuclear wastes, and as gentle as a hair-thin strand in diagnostic probe of an ailing human body. The service of glass has been tailored for the times over thousands of years, and that is not about to change. Glass scientists with vision have been preparing for the new millennium, and there can only be a reconfirmation of the union of sand and fire as a miracle of chemistry.

Chapter Eleven

New Glasses for a
New Millennium

Little more than a hundred years has passed since we first tapped into the versatility of glass for scientific purposes, and yet here it is at the dawning of a new millennium, deeply rooted as an indispensable material for the betterment of life, and fully vested with the promise of even better things to come.

The gifts of glass have been many, and scientists are expecting more—much more. Certainly, the material has shown that it doesn't outlive its usefulness, but the rate of progress for new developments in the times ahead depends to a large degree on the willingness of industries and universities to invest in long-haul research with no assurance of financial reward. It matters that these are the days of the long knives, with their bold slashings of budgets and staffs, and lab work on speculation.

Also, a broader understanding of the structure of glass will be necessary if scientists are to determine how far the limits of the material can be pushed. If glass can be made pliant enough to take stress without fracturing, all the while leaving its inherent Herculean strength undiminished and retaining its resistance to heat, then the union of sand and fire will have a triumphant new meaning.

There will be glasses with semiconducting capabilities, and glasses with high resistance to fire. The development of scintillation glasses, to be used for detecting charged particles, is likely to occur, as are advancements in the fabrication of glasses that can emit laser light (for a glass to function as a laser, the atoms must be excited by an external source, such as a xenon flash lamp. With the energy level of some of the electrons raised by the illumination, it follows that when the level drops back to normal, the excess energy is released—and radiated as laser light. But for this to happen, the glass must contain certain rare-earth elements such as erbium and neodymium).

The many possible uses of membrane technology involving porous glasses are all too alluring to be ignored, especially when the size of the pores can be controlled. Glasses with such powers of precision filtration shine with the potential for use in medical procedures. Indeed the enlistment of glass in the cause of healing can only take on added importance in the coming years as the material is made to be even more biocompatible, and as the glass fiber optics are threaded on even bolder courses through the human body.

Fiber optics, as much and maybe more than any other application of glass, will carry the material strongly into the future, not only in the fields of medicine and biotechnology, but in telecommunications. Even now, the land is being laced with glass fiber, and in laboratories at glassworks and universities, scientists are feasting on the versatility of the glassy thread.

For Corning, the manufacturing and marketing of fiber-optic glass brings in the most revenue today, and with a projected increase in demand for the material of 25 percent annually for as long as it takes to replace all the copper wiring in the ground, the boom seems certain to continue for some years. Corning controls at least half of the market for glass fiber optics in the United States, and that, more than anything else, has served in recent years to position the company on firm financial ground.

Glass fiber optics will tie the world together for life in the new age of rapid communications, when advanced concepts of voice and information exchange become commonplace. And there will be other important technologies woven with that thin thread of glass, such as the one that will have us spinning off into worlds of spectacular illusion.

Called virtual reality, it promises to deliver the means by which information is moved not over a keypad, but by three-dimensional imagery. With this technology, a real estate agent will preview a property for a potential buyer by having him wear glasses on which three-dimensional views of both the exterior and interior of the house are projected with the use of fiber optics. It is reality to the degree that the viewer will likely duck when descending through the low overhead of the cellar stairs, and maybe even reach out to feel the smoothness of the marble counters in the bathroom.

Three-dimensional technology requires a lot of bandwidth, or information carrying capacity, and glass fiber stands above all else in its ability to meet that demand. It is not a new technology, of course, since movie-goers as far back as the 1940s were clutching the armrests of the seats as snarling tigers leaped at them from the silver screen. The magic of that terrifying illusion was embodied in the cardboard and plastic glasses handed out to the patrons as they entered the theater. Had they chosen to look at the screen without the glasses, the image would have been blurred and the tiger's fury lost in uncertain focus. The new virtual reality is much more complex in that it can allow the viewer to participate in the illusion—to slay the tiger, in effect.

Virtual reality will take many forms, incorporating a vast variety of devices. It has already begun, with simulators for flying and driving, each with reality so stark as to have the participant almost feel the rush of wind, or be sickened by turbulence at eighteen thousand feet. And if the virtual-reality games in existence today can be compared to Pong and other early video games more than twenty-five years ago, then the next few decades should bring a similar leap in game technology. The joystick will plunge our children and theirs even further into the action, onto a playing field no longer flat and featureless.

But there is another side to this. The same joystick that drives the victor of good over evil in a computer game is capable of simulating serious pursuits, such as that of an oceanographer on a three-dimensional dive aimed at determining how best to examine the coral of a reef. In the same manner, firemen in training can make their way through the chokeless smoke and unsingeing fires of virtual reality to perform their meritorious rescues. Indeed, the possible uses of this technology are far-reaching, since, beyond simply projecting images, this is the creation of environments, dimensions pulsing to life through a thin and flawless glass fiber.

Glass fiber optics, by all present indications, will be an essential material whenever more advanced methods of moving information and communications are brought about. It is clearly the current best way to transport a voice from one telephone to another, or to feed instantaneous updates of stock market quotations to the computers of investors. It is a glass of brilliant fulfillment in the electronic revolution, and it is not likely to diminish in value when science moves on to something even better: the photon, as a component of information storage and retrieval systems.

As Dr. Donald Keck, one of the primary developers of

glass fiber optics, told me as far back as the early 1990s, "Without question, photons are the wave of the future." His perception remains valid, as processing information with photons, or light, rather than electrons, is set to come about as the next major change in the evolution of computer science. Optical technology using the glass fiber optic as a major component will make a profound difference in the handling of computerized data.

Massive amounts of information can be processed in record time using an optical computer, but to do that there will be a need for a carrier with a high-load capacity, and one of near flawless performance capability. As of now, only glass fiber optics can provide all of that.

Photonics is one of the eight core technologies that Corning will be focusing on in the coming years. Another will be glass-ceramic, the material invested with super strength through partial crystallization. Even stronger glass-ceramics will be in the offing, and with them, renewed interest in the often dismissed but never forgotten possibility of using glass components in the engines of automobiles. The attraction is always there: with glass, there would be little heat conduction, and therefore no need for cooling or lubrication. It will work most efficiently with a diesel engine, but with improved glass-ceramics, who cannot say that the day is coming when a gasoline engine will purr with the lambent resonance of crystal?

Bold dreamers, though, envision something even more grand. Looking to the seas, they tell us that extensive use of glass in new-era oceanography cannot but be ordained. Glass can take the depths where steel would crumble like tinfoil. Glass can take the full fury of the sea, as witness the fishing-net float that comes ashore virtually intact after years—maybe ten or more—at sea, years when it was borne on storm-stirred swells, teased by sharks and whales, en-crusted with coral, scarred by the wind.

There is cause, then, to exclaim over finding a glass float on the beach. Only a small percentage of them make it to shore. Most come from Japan, and the chances of recovering a float are best in the outlying islands of the Pacific, although the coasts of the western states, especially Oregon, can be productive. Collectors of the spheres prize their hoards, for there is a strong sense of mystery about them. It's enough to just sit and look at a float to have the sea rush in and flood the mind with fantasies of gale-whipped sail, and Oceanids who cavort during the calm that follows.

The floats for the most part are made of thick glass, and quite often there will be a little water inside the device when it is found. The seepage occurs through the point where the glass was attached to the blowpipe or automatic blowing mechanism. Nothing is so basic to the glassblower as the crafting of a sphere. The breath inflates the gather of molten glass as it does a balloon. But unlike a balloon, the glass will sag and become distorted; it cannot be pinched off with the fingers to hold the air in. It is essential that the glassmaker maintain a timely rhythm with his breath and hand motions if the glass is to form a true globe.

However, in the making of floats, perfection is not a de-manding priority; it is one of the few times when glass is allowed to become slovenly, but with a fishing float, does it really matter? I put the question once to a glassmaker in Japan, where the best floats are made, and this was his reply: "What do yellowfin tuna know?"

Let me cite one thing that yellowfin tuna do not know. The fish do not know that if a round piece of glass the size of a volleyball stays in the water for years and years, floating seven thousand miles, and finally comes ashore little worse for the wear, there will come a time (mark this) when the floor of the ocean's deepest trench, more than thirty-five thousand feet down, will be reached by a vessel made of glass. Marine scientists will board spheres of glass and descend to those forbidding depths without fear of being compressed to death.

Glass capsules to lift devices that take samples from the ocean floor are already in use at depths of twenty-three thousand feet, and even that is beyond the limit of resistance for steel. Of course, thicker metal can be used, but then the buoyancy might be lost. And of course no matter how metal is restructured for use at killer depths, the transparency of glass cannot be duplicated.

In a foreword in a book about some of the works of glass artist Dale Chihuly, Sylvia Earle, a renowned oceanographer, writes of the importance of that transparency: "My dream, I told Chihuly, is to have a glass submarine, a clear sphere within which I can sit, warm and dry, and fly to the ocean's greatest depths, seven miles down. . . . Glass . . . is transparent, a vital consideration for those such as I who like to see where we're going underwater."

If something made of a material other than glass does make it to the floor of the Marianas Trench in the Pacific, off Mindanao, it will most likely be a composite in which the strengths of two or more materials have been combined—the strength and heat resistance of glass, say, and the toughness and workability of a plastic. That indeed would be chamber music by otherwise discordant strings of molecules.

The science of materials has been plentifully productive over the past century, especially since the end of World War II. And the scientists, materialists in the true sense of the word, are still chasing the elusive composition that will draw all the strengths that chemistry has to offer, and yet be lightweight and amenable to a wide spectrum of shaping and form. Glass drawn into fibers has come as close as anything.

At Corning, experiments have been conducted in which glass and plastic—70 percent of the former, 30 percent plastic—have been melted and processed together. The plastic is one that becomes molten at the same temperature needed to melt the glass, and therefore they're intimately mixed together.

"The properties we end up with," said a Corning official, "are midway between those that you have in a plastic and those that you have in a glass. Glass is very brittle, and it breaks and shatters easily. This material containing both glass and plastic is not brittle; it has picked up some of the toughness of the polymer. On the other hand, polymers have flex. But when you put all this glass in there, it doesn't flex anymore.

"Tough as plastic, but without flex, like glass. That's a material for a lot of things."

In addition, the glass-polymer material will not burn, not even in an atmosphere that is 100 percent oxygen. "So imagine the interior of an airplane when it's of critical importance to have the least flammability that you can, and still have material in there that looks like plastic, that's formable and can be made in sheets," the man at Corning said. "The best polymer available today will burn in about 40 percent oxygen. So if you get in a crash, a flashover, where you get a lot of high-temperature flame, and the polymer starts to vaporize, it will burn. This combination of glass and plastic will not fail like that."

Also in the future are "glassy" metals. These alloys are the subject of heavy research, and if scientists can refine the technology for making them, material science will have recorded an important achievement.

But is it glass?

Basically, the idea is to get the atoms of a metal to fall into disarray, as do those of glass. To do that, the melt must be cooled in such a way that will thwart the tendency of the molecular structure of metal to quickly put itself in order and crystallize. Immersing in a cold bath—a flash cooling—will make the metal melt glassy, but only droplets can be processed that way. Something more promising has been found, however, and it involves using a blend of different materials in the mix for melting. With different atoms of different sizes, the rush to order encounters confusion, collision, and delay enough to allow the disorder to prevail in the cooled, solid state.

Such a material offers several advantages. It is considerably stronger than it would be in its natural crystalline condition, and it possesses only half its natural density, meaning it is lighter. And at least a dozen other glasslike properties are present in the glassy metal, such as the ability to resist corrosion. (Mother of Mercy, is this the end of rust?)

The success of this technology largely depends on cooling speed and the mix for the metal. Allied-Signal Corporation, in Morristown, New Jersey, has succeeded in producing an iron-based glassy metal with the use of boron in the mix. The material, with superior magnetic qualities, is used mostly to make distribution transformers for electric utilities.

But there are major limitations to the use of such amorphous materials. They can only be produced in powder form, or as wire or foil. To attempt to make ingots or I-beams of them would mean they would have to be subjected to high heat after having gone through transformation to a glassy state. Such heat will cause these metals to recrystallize, and thus forfeit all the advantages of sharing the molecular disarray of glass.

Scientists may find a way to overcome that drawback. Meanwhile, one can argue that the reputation of glass will suffer because of this infringement on its good name by the likes of beryllium. Will future editorial tinkerings with the words of the Bible have it being said in Revelations, "And I saw as it were a sea of nickel mingled with fire"? Most important, it would not do, I think, to call something glass if you can't see through it. That's the thing—the transparency; if it's glass, it has to let you know what's on the other side, and metal can't do that, no matter how the molecules are structured.

While the new amorphous metals will not make it as windowpanes, research in progress has targeted flat glass used for windows, including that for use in automobiles and airplanes, for even smarter capabilities. Most of all, it will involve the introduction of new coatings that will give the glass more control over comfort features, such as temperature and light, in houses and office buildings. For example, the coating may function as a venetian blind, opening and closing as needed. Such technology is already being used in many office buildings, but it is not yet so advanced as to satisfy the tireless probings of glass scientists.

At the least, a smart building of the future will use the venetian-blind coating technology not only to regulate the quantity of sunlight entering a room, but, by opening and closing the coated "slats" (microscopic lines in the coating), to direct it to the ceiling, from where it is reflected to windowless cubicles normally denied sunlight.

Other coatings will further reduce reflection on the television screen, and for the merchandizing of a product, improved nonreflecting glasses will showcase the item with flattering clarity. Here is Denmark's Bang & Olufsen, maker of high-end, high-style audio equipment, coming out with a piece for the home entertainment center that is both powerful and pleasing, as is the way with Danish craftsmen. It is a CD changer with six disks laid out adjacent to one another, and all under a cover of clear glass.

The visibility factor—the glass—clearly lifts this piece into a new realm of music appreciation. To look at a compact disc when it is playing (and to look to see what comes next) is, for many, to give the sound more body, to give the listener an added edge of comprehension. The changer can be positioned at least a half dozen ways, including hanging it on the wall like a Chinese scroll. The visibility factor has a price, though: $4,000, speakers extra.

The use of glass in home decor may be on the road to trendiness again. Glass doorknobs are coming back into favor, as are coffee tables and other tables made of decorative glass. Interior walls of glass brick are being touted by decorators and architects, and there are even reports of indications of a forthcoming resurgence in the use of glass globes in the garden: Out of the Victorian era, a globe of colored glass is meant to rest on a pedestal of cement, as having been enthroned for sovereignty over the flowerful realm.

What will not be revived is the dying picture window. Somehow, the figure of 100 billion square feet has surfaced as the area covered by all the windows in the world, and that number can only diminish now that the picture window is being rendered out of designs for new houses. Michael Pollan, writing for the magazine *House & Garden,* has noted that the window has even been outlawed in the Florida panhandle town of Seaside, on the Gulf Coast.

"The unfenced front lawn and the unmullioned picture window were flags of allegiance to the suburban codes of like-mindedness and nuclear normalcy," Pollan reminded us. "The family that flew them both were saying that it had nothing to hide from its neighbors, that it led an inspection-worthy existence."

But then, after all, there will be no lessening of the need for window glass because of the demise of the picture window, for, as Pollan writes, "I noticed that the picture window of my childhood has been replaced by something much more fancy that I later learned is called a 'Signature Window': a dizzyingly tall pileup of windowpanes of every conceivable shape and arrangement, typically installed above the front door of a house and invariably crowned by a grandiose semicircle of glass."

Glass. It sticks to our needs like a prayer.

PART III

THE GENTLE SIDE

*Glass is more gentle, graceful, and noble than any metal,
and its use is more delightful, polite, and slightly than
any other material at this day known to the world.*

—ANTONIO NERI, 1612

Chapter Twelve

Ascending Out of Craft

Not only in technology and commerce has glass widened its horizons. The use of glass as art, a tradition going back to Roman times, is also surging. In Seattle and in the mountains of western North Carolina and the countrified south of New Jersey—nearly everywhere, it seems—men and women are blowing glass and creating works of art. In recent years the movement has gained new status; it has emerged, and how, from the hobbyfied depths of craft.

There are artists blowing glass today who possess immense talent, and they are responsible for the new and trendy glorification of the material. Suddenly (it seems that way), a stumpy, affable meat cutter's son from Spokane has become one of the most financially and critically successful artists of the twentieth century. The things Dale Chihuly does with glass give new meaning to the miracle of sand, soda ash, and lime boiled together. And in just twenty-five years, under his swirling influence, glass has become the medium for a major force of art.

Because of the need for a furnace, glass artists for many years were denied the freedom and convenience of working in a studio. Instead, they worked, when permitted to do so, at a factory where commercial glass was made. Even the great works of glassmen such as Tiffany, Lalique, and Émile Gallé, the first glass artist of modern times to sign his pieces, were done in the workshops of factories, albeit their own factories in the case of Tiffany and Lalique. And, in truth, their works were more high craft than art for art's sake. It wasn't until 1962 that the modern glass art studio movement began.

In that year, Harvey Littleton, a professor of fine arts at the University of Wisconsin and son of a former director of research for Corning Glass, discovered that glass could be melted at temperatures low enough to permit use of a small furnace. It meant that glass for art could be blown in a home studio. Also figuring prominently in the introduction of studio glassblowing was Dominick Labino, a physicist with technical genius, who died in 1987. He and and Littleton would come to be regarded not only as founders of the studio movement, but also as two of the finest glass artists in the world.

Littleton established a workshop on the grounds of the Toledo Museum of Art, and for the first time, students were invited to explore glass outside of the factory atmosphere. With that, the making of glass as decoration began to pull away from industrialized domination. The steamy celebration of the making of art glass would begin soon after, and continue to this day. Some of the resulting art is of resounding artistic quality, a lot of it is not. Much of the bad is overwrought, or sparse with a simplicity that has slipped into feebleness.

By making it possible for a glassmaker to work independently, on his or her own, Littleton and Labino completely rewrote the script for making decorative glass not only in the United States, but in much of the world. Suddenly, there could be soloists, and for all but the weak voices of the glassmakers' chorus, it was sweet.

The Metropolitan Museum of Art in New York waited until 1996 to mount its first exhibition of conceptual glass art. Writing in the catalog, Jane Adlin, cocurator, said, "The studio glass movement has become an integral part of the art world, and the artists' ongoing explorations of new forms, new combinations of materials, and new ideas promise an exciting future. The Metropolitan Museum of Art continues to collect the most significant expressions of this fragile yet powerful medium." When asked if the museum had dismissed all apprehensions about glass as craft rather than art, she replied, "Absolutely. We at the Museum have always felt—or felt for the past twenty years, anyway—that it was just a matter of timing when an exhibit could be mounted."

American artists have responded to their emancipation from the factory workshop with cyclonic exuberance. There are numerous schools offering instruction in art glassworking now, when there were few if any before the onset of the studio movement. In addition, programs involved with glass as a medium for art have been added to the curricula of some colleges and universities. And the movement in its early years benefited from an expanding art market in general. Times were good, and with more money, more people turned to collecting, to satisfying the inborn urge among many to gather things of craft. And glass would prove to be equally alluring to more sophisticated collectors. In Europe and Japan, too, the making of glass art has drawn heavy attention. But the movement is centered in the United States, and here most of its best artists work.

"Glass in this country is a whole new world in terms of being a medium for art," said William Travers, proprietor of an art gallery in Seattle. "Twenty years ago, in the 1970s, when I first started showing glass art, no one was buying. People had not seen glass art before, and they had no idea what they were looking at. Now there's a lot of interest in it. Glass is so old, and culturally, I think, we are calling out for these things from the past. And there are other factors, such as the intensity of color that can be achieved only in glass. The artists? The number of truly great ones can be counted on the fingers of one hand."

On a shelf behind the desk where Travers sits are two glass objects, a tall pitcher and a smaller vessel. The glass is red with black trim, but this red is truly unique and probably could not have been obtained with the use of any medium other than glass. In one respect, it is a soft red, like that of rain-washed bougainvillea, and then again it is a red to signify passion and to caution against the humdrum. The two pieces appear in arresting symmetry, and as a strong definition of grace in a material. It is an example of one of the better pieces of glass art being made, and its creator, a resident of Seattle named Dante Marioni, is among the glass artists Travers would count as great.

Marioni has been blowing glass for more than thirty-three years. He is a superb technician, a person for whom a gather on molten glass on the end of a pipe is a seed of endless creation. His working of the glass is masterly, and when he talks about his art, he talks as a glassblower and not a designer.

Seattle is the center of the studio glass art movement in the United States. As a whole, glass artists are of a type who prefer an informal lifestyle, with easy access to the back roads. While Seattle bears the standard of urban chic in the Northwest, there is still a bit of manure on the sole of its shoe, and that suits many of the artists just fine. When painters gather, they seldom speak of the merits of a certain brush or the absorption quality of the canvas, but with many glass artists, the subject of technique drives their talk. Then too, there is a strong affinity between Seattle and trendiness (the countless coffee shops, the thick aura of computer-age superiority), and in the art world today, glass as a trendsetter is a runaway favorite.

And finally, there is a school for glass artists not too far from Seattle, a place called Pilchuck. It is the font from which springs the genius of the studio movement. Consider William Morris.

He's known to all of his close friends as Billy, and he rides a motorcycle, not just around Seattle, but all the West, and for six months each year he is at the furnaces at Pilchuck, blowing glass and fashioning art pieces that have won high praise throughout the world. He came out of California and got a job at Pilchuck as a truck driver; he grew up with that school, progressing from apprentice to master, and then on to stardom. Two of his works were included in the exhibition at the Metropolitan Museum of Art.

Of that exhibit and the artist, Jane Adlin said: "Bill Morris is working on the leading edge of what's being done in glass. In terms of artistic content, his work is not matched by anything else that I can see."

Morris, who turned forty in 1997, has followed well the path of opportunity opened up by Harvey Littleton and Dominick Labino in the 1960s, guiding himself to a unique expression of art in glass. His main interest is in the primitive—its myths and rituals, and the artifacts left behind. He is a hunter, but with a bow, and it is the ritual of the exercise that attracts him. In 1988, he began a series of works focused on artifacts, such as a tusk, a primitive tool, or a gourd, often in suspended arrangement. They were all highly effective, despite being opaque and dense for the most part, and therefore in defiance of the aesthetically correct play of light in glass; still, some translucence is retained, and with properly placed artificial lighting, there is a serene, faintly elusive inner glow to be had.

His other well-known series is of glass casts of canopic urns, in which Egyptians of ancient times put the viscera of the dead to rest. The jars are lidded with representations of animal heads, and the vessels are embellished with drawings. To further appreciate the artistry of these pieces, it helps to know that the heads have been fabricated not in a mold, but by hand, tweezing molten glass until there can be no question that the masterpiece of a head on the urn is that of, say, a coyote. The translucency of these jars is such that the passage of light through them is like a resurrection for the dead they represent.

Much has been written about Morris's work, and quite often it is so laudatory as to trivialize the earthliness of the man. They write of the work's spirituality, and (one of my favorites) even of the way the sculptures "capture the corporeal insults that follow the cessation of life." What really happens most often when a person sees a Morris piece for the first time is this: He or she first expresses astonishment that the material is glass, and then judgment is passed on its quality as art. It is very difficult to overshadow the concept of glass as a material.

"I like the idea of pushing the limits of glass," Morris said, "and there is still a lot to be done with it. But that always takes a backseat to the content." But with glass, it all goes together, and dispossessed of his enormous technical ability, this artist would be but a wounded lion in the studio glass art movement.

Although Morris does his work at Pilchuck during the winter months when the school is closed, he has a studio not far away, down a rural road and up a gentle hill, and then onto one of those tauntingly (because you're not able to pack up and move there) pleasant homesited swells of trees and rhododendrons found throughout the Northwest. In the nicely finished loft of a barn there, there are a number of his pieces, and they do indeed invest that place with a pervading spirit of genius. For the price of a new Lexus, a William Morris work waits in the barn to be taken home, there to be positioned best in a strong room, such as the study of an adventurer.

Morris's work in the hot shop, with a team of assistants, is artistry in itself. In observing many such talented glass-blowers at work, I was twice driven by an urge to create my own signature piece—nothing grand, but a modest vase with, I hoped, a certain lyrical quality.

Dawn came frozen and hushed to the late winter's day, but it was a good warmth that filled the hall at the great Swedish glassworks, Orrefors. It was here I had come to make my vase. It started well enough with words from an old Hindi saying dancing in my mind: "If you are a blower of glass, fashion the cup as if it were to be touched by the lips of your beloved."

One end of the five-foot-long pipe for blowing was thrust into a furnace through the glory hole and twirled around to collect a gob of molten glass, much as a fork is twirled to gather spaghetti. At a temperature of 2100°F, the mixture of sand, soda and lime was rude with glaring color, and thick and inching.

Earlier I had watched Juhani Karppinen, an employee at Orrefors for more than twenty years and now a gaffer, or master glassblower, at work, and marveled at the way he coaxed form from the molten glass. "You will know by the feel of it on the pipe if it is right," he told me.

Astonishingly, a vase began to take crude shape as I blew into the pipe. I stood atop a box holding the pipe so that it pointed straight down, allowing the blown liquid to swell at the base and pull down to form the start of a neck.

There was more blowing until the glass drew thin and the neck long enough to take the stem of a dahlia. The men assisting me pronounced it done and smiled to signal that it was a vase of some distinction.

It was only after retrieving the piece the next day from the annealing oven, where it was placed to cool over a period of four hours, that I discovered the flaw: When put down, the vase tended to roll from side to side, like a bottle set adrift in the surf.

At another time and another place I was invited to blow a Christmas-tree ornament, and that was successful only to the point where the piece can be hung at the back of the tree, along with the flotsam of arts and crafts from summer camps.

At the Studio, the new glassworking teaching facility of the Corning Museum of Glass, summer classes in glassblowing and other aspects of working with the material attract people from throughout the United States as well as many foreign countries. For a typical session of glassblowing instruction, there will be two or three students intent on becoming successful full-time glass artists. Most of those who enroll, however, have little desire to raise their glassworking above the hobby level.

"I started working with glass when a store wanted twelve hundred dollars for a piece I was interested in," said Ron George, a student at the Studio. "I said, 'Hmm, I think I can learn to do that,' and that's what I have been doing for eleven years. Now that I'm retired, I want to get into it more."

Then it was his turn to fashion a flower from molten glass, and he did it well, tweezing the petals into being and pulling the viscous glass until there was a stem. It was something worth keeping, but this was school with the promise of better things to come. The flower went crashing into the box holding cullet. For some, like Ron George, meeting the challenge of the hot shop is a pleasurable accomplishment. Some others are cowed at the start by molten glass and are never able to take command.

Students who take instructions at Corning have the added advantage of being within walking distance of the finest and most extensive collection of glass in the world. The Corning Museum of Glass, where the evolution of glass is chronicled in exhibits spread through seven galleries, opened in 1951 and has been getting better ever since. It covers the full journey of the material through the ages: a perfume bottle in the Egypt of thirty-five hundred years ago, a piece of art blown in a Seattle studio in the 1990s.

There is a rare Roman cage cup, or diatretum, from the fourth century A.D., and a stunning leaded glass landscape window by Louis Comfort Tiffany. Those are two pieces in the collection. There are twenty thousand more. And along with the glass, there are some of the printed words of those who made it—this, for example, from a letter by a German engraver named Friedrich Winter in 1686: "Glass engravers and also cutters are so numerous that the one spoils it for the other and hardly anyone can earn a proper living, and each should teach two or three youngsters, but when they have brought it so far as to be able to do a bit of scratching, they run away from their masters and set up with their own tools and bungle away each one for himself."

That doesn't seem to be the case in a shop in the museum complex where Steuben engravers do their work (Corning owns Steuben). It would do here to retire the myth about the passing down of talent from engraver to his son. "Of all the engravers who have worked at Steuben, none had a father who was also an engraver," Roger Selander, an engraver, said, sitting at his copper wheel in the small studio where he works with cold glass, putting an eagle to flight on its surface, or a ship under sail in its glistening sea of reflected light.

For artists who work with glass when it is cold, there is no second chance, no reaching through the glory hole of a furnace for a new gather of the molten material. They are the engravers and etchers and cutters, men and women like Selander, a master, and Eamonn Hartley of Waterford, another master who can put the face of Noah on a piece of crystal and give it all the authority of a Biblical injunction.

Long before there were copper wheels and diamond points, glass was being engraved with crude, sharp tools. In the fourth and fifth centuries, many Damascenes no doubt took their saffron-yellowed rice from bowls scrolled with engravings. And it was all done, and done well, by hand. Much later, in Germany and pre-partitioned Czechoslovakia and some other parts of Europe—lands where there was little or no hot working of glass as art at the time—engraving and other techniques involving cold glass were raised to new levels of excellence.

In America, foreign-born engravers were teaching their art to apprentices, and that chain of handed-down skill reaches to the present time. Sculptors and other artists were being hired to create designs for engraving. In Corning, Steuben was founded in 1903 by the Englishman Frederick Carder, who, over a period of thirty years, guided the firm to its position as a glassmaker of wide renown. Under Corning, Steuben has continued to excel as a provider of high-grade crystal, much of it engraved. Started in 1947, it is a tradition in Washington now that Steuben crystal be presented as gifts of state. Great Britain's Queen Elizabeth, for example, on a state visit to the United States when George Bush was president, received Steuben's "Shakespeare's Flowers," an engraved bowl. For President Anwar Sadat of Egypt and Israeli prime minister Menachem Begin, there were sparkling arcus-shaped pieces presented by President Jimmy Carter.

Some Steuben pieces are still popular and still in production more than 50 years after being introduced. One of them is called the "Gazelle Bowl," a piece standing a little more than half a foot high and about the same in diameter, and engraved with a frieze of twelve leaping gazelles. An art deco work first made in 1935, it was the first of the firm's designs using heavy engraving.

Roger Selander has engraved the "Gazelle Bowl" on several occasions when it was ordered on commission. In doing this, he worked in the ages-old manner of the glass engraver, sitting at a small lathe and working the crystal against the copper wheel revolving there; he would spend about three hundred hours on this one commission. When finished, the engraving appears as if in low relief, although it is depressed below the surface.

Engraving on crystal is difficult, exacting work, and that, of course, is reflected in the prices. Masterworks such as the "Gazelle Bowl" cost tens of thousands of dollars. People of means—collectors, especially—continue to buy the works, and sometimes the cost can go well into six figures. "I worked on one engraving for eight hundred hours," Selander said, "and it sold for two hundred seventy thousand dollars. The theme of the piece was 'Romance of the Rose.' " I later learned that the owner is European and had made a fortune selling plastic eating utensils to airlines.

Cold-glass sculpture is exuberantly caught up in the current sweep of popularity for glass art. Peter Aldridge, a former designer for Steuben, is among the best doing the work. He applies computer technology to his art, resulting in spellbinding abstract art of severe precision. With the help of the computer in controlling the sculpting machinery, such as a diamond-impregnated wheel, he can achieve the sharp angles and smoothness that play so well with light.

Czeslaw Zuber, on the other hand, goes at his glass with tools more suited for carpentry or cinderblock work. His wonderful depictions of the heads of beastly creatures are fabricated not from good crystal of flawless clarity, but from blocks of industrial glass. He takes a hammer to that glass and bangs his way to a rough representation of what he wants before turning to sandblasting tools for closer refinement. And with all of that, this Polish artist who resides in Paris produces glass art of museum quality.

Of all the crystal-based glass art, none gives off such a lordly aura as the chandelier. Given a proper hanging, a crystal chandelier can lend nobility to a room as it has done in the castles of kings and the summer palaces of tsars. Alas, the days of such grand illumination, when the crystal is arranged on the metal frame in such a way as to have a perfect breaking of the light, appear to be fading fast because, if for no other reason, the ceilings of today's houses are too low to accommodate such hangings. And even if the height is there, the room must have ample width and depth if the chandelier is to be effective. It just will not do to hang a forty-eight-light, $99,400 Baccarat crystal chandelier where it will throw shadows on a wall or can be cleaned without the use of a tall ladder.

Many of the great old chandeliers, such as the one in the East Room of the White House, are Czechoslovakian-made, often with glass arms, crystal bowls, and tiers of candles and pendeloques. And in Vienna there is a name of high honor in the making of crystal chandeliers. Now in its 175th year and fifth generation of family-run ownership, J. & L. Lobmeyr counts among its masterpieces the massive forty-foot-wide chandelier now inside the Hall of the Supreme Soviet, in the Kremlin, and another made in 1966 for the Metropolitan Opera in New York.

It is right that Lobmeyr, with its old-world reserve, be in Vienna. Until recent years, the capital of Austria was the city, of all cities in the world, best suited for the grandness of crystal chandeliers. They simply belonged there, throwing down stylish light on the Sacher tortes and the uncertain smiles of bankrupt royals.

"If you are talking about retaining traditional techniques for making chandeliers, it is best to have a small firm," said Harold Roth, a member of the firm's controlling family. "With us, we still like to have mouth-blown glass for our chandeliers. At the same time, we are very open to new contemporary styles in chandeliers."

He is in his office on one of the five floors of the building where crystal and stemware and other high-quality glass are displayed, and he is telling about the chandeliers made for King Gustav VI of Sweden, Queen Isabella of Spain, and others, including "a Bulgarian king. I can't remember his name." What is most conspicuous there is a large chandelier that hangs in the main showroom. It weighs close to half a ton and was made for an Arab of unusual wealth. It was to have hung in the reception hall of his palatial house, but when it was finished he had second thoughts. So it remains at Lobmeyr, its fate uncertain, more dust collecting on its drop prisms, its graduated chains of crystal beads, and the clear bulbs of its lamps.

I asked Harold Roth this: "How do you clean a chandelier of that size?" And his reply was, "I have no idea."

At Orrefors, Sweden's premier maker of glass art, the emphasis had been on design since production began there in 1898. As far back as 1917, the firm hired two artists, Simon Gate and Edvard Hald, both painters, to supervise not only the design operation, but all production at the *glasbruk*. It was one of the first departures from the tradition of the house's master blower being in charge, and it served to introduce fresh and vital creativity in the making of decorative glass.

Examples of Orrefors chandeliers can be seen in the Grand Foyer of the Kennedy Center, in Washington—where there are eighteen of them. Another, weighing two tons, hangs in the General Motors building on Fifth Avenue in New York. All of those were designed by Carl Fagerlund, a lighting architect who was with Orrefors for nearly forty years.

The Orrefors designer and the glassblower work together to first determine if the proposed design can indeed be fabricated, and if it can, to then see it through no matter how burdened with problems, including clashes of ego. Making glass art at Orrefors can cover a wide span of esoteric techniques. They engrave, of course, but so do they use graal and ariel, two techniques in which surfaces of clear glass are applied to cores of colored glass into which the designs have been cut, engraved, etched, or sandblasted. In another process to make what is called Ravenna glass (after the Italian city of mosaic fame), a design is sandblasted into a disk of layered glass, after which crushed colored glass is used to fill the depressions. After being reheated and worked into final shape, the piece becomes like a mosaic. All were developed at Orrefors.

There will come a time, Orrefors's Helen Krantz believes, when laser technology for making glass art will come into use. Not only could the tool cut and engrave the crystal, but it could supply the heat for melting as well. As a Scandinavian, she is in love with design, and as a designer, she is equally in love with glass.

For the most part, glass art such as that from Orrefors is meant for display—a shelf or pedestal in the light—and probably never again will it be used as ornaments for the hoods of automobiles. Early in the twentieth century, jeweler-turned-glassmaker René Lalique, a craftsman of highest skill, *maître verrier,* began fabricating the hood ornaments, mostly in art deco style. They included animals and human heads, and almost always they were posed as facing down the wind with hair billowing and ears pressed flat against the head. Sometimes they were illuminated with a tinted uplight.

There is no prouder name in the history of decorative glass than that of Lalique, and pride is what the driver must have felt with one of the Frenchman's crystal eagles perched on the broad, black hood of his Pierce Arrow.

Europe was building on its tradition in glass art through the late 1940s and all of the 1950s. The resurgence of Venetian glassmaking that began in the early 1920s had carried over. In Finland, where there has long been a love for design, glass art rose to new heights of brilliance. And all of this was achieved at the factories, not so much by single artists, but through the joint efforts of designers and gaffers, or master glassblowers. But they were little known. It was the company's name that graced the work.

Two of the Central Europeans working in glass during those years have emerged as priest and priestess of contemporary glass art not only in their native Czechoslovakia, but throughout the world. They are Stanislav Libenský and his wife, Jaroslava Brychtová, both of Prague. There can be no question that they are a strong force among the artists for whom fire is a palette. The art they do in collaboration is large and powerful: glass beams, crystal walls, great sculptures of red glass that glow like the sunset sky. For the opening of the new Corning Museum of Glass building in 1980, Libenský and Brychtová were commissioned to do a piece for the entrance area. *Meteor, Flower, Bird* is cast in three parts and with its grace and muscle, it both mocks and confirms the gentleness of the material.

Meteor, Flower, Bird is not an installation for the home, but only because it is seven feet high and seven wide. Fittingly, it is in the company of the world's greatest collection. Also at the museum is a monumental Libenský-Brychtová work entitled *Green Eye of the Pyramid.* In this, a geometrization gone mellow in light, two pyramids merge at a juncture with an optical effect of a glowing eye. To make this difficult and somewhat anguished work, Libenský and Brychtová made melts of glass in five molds.

Other glass artists openly marvel at the couple's work, and when it is written about, the praise is effusive, and deservedly so. Susanne K. Frantz, curator of twentieth-century glass at the Corning Museum, has edited a book on the two Czech artists, and in it she writes this: "The beauty of their art is always subservient to their ideas. Many artists grapple with these ideas: however, Libenský and Brychtová choose to address them without artifice or illusion. As superb craftspersons and technologists, they have helped push the material toward its maximum potential as an architectural and sculptural medium."

Maurine Littleton, Harvey Littleton's daughter, owns an art gallery in the Georgetown section of Washington, D.C. The works of many of the best contemporary glass artists are exhibited there, such as those of Erwin Eisch, a German who uses glass with little regard to the interaction of light. Early on, he refused to be guided in his work by so-called standards of good design in German glass art. Whether a glass telephone painted gold or a bust of Picasso with drawings on his cheeks and chest, Eisch's pieces speak strongly of individuality of expression.

And Mark Peiser. Littleton had a piece of his called *Mountain Skyscape* in her gallery, and it was dotting that room on Wisconsin Avenue with jewels of light. Later, I went to meet the artist in his studio on top of a mountain in western North Carolina. That state, like the Pacific Northwest, has attracted glass artists for a number of years (Harvey Littleton continues to live in North Carolina), and again, that has been mainly because of a school founded in the infancy of the studio movement. The Penland School of Crafts, in the mountain town of Penland, is not as art-conscious as the better-known Pilchuck near Seattle, and, unlike Pilchuck, the instruction is not limited to glassmaking.

But Penland has an honorable background in fostering the use of glass as a medium for art, and, as the school's first artist in residence with a specialty of working hot glass, Mark Peiser has been an important part of that.

Peiser does not refer to himself as an artist. In fact, he is a genius when it comes to the technology of glassmaking, but he is also a superb designer. He just doesn't know if that adds up to art, and, in truth, he doesn't much care. In fact, Mark Peiser isn't certain he cares much about glass anymore.

He was in the studio movement from close to the start, and there was no contemporary American glass art more avidly sought than his works of the 1970s. But now he seems to be beset with a loss of faith in the integrity of what has happened since Harvey Littleton and Dominick Labino took glass art out of the factory and put it into the studio. It is important, he feels, that the glassworker master technique, that he or she have the formulas for glass melts stored in the mind and ready for retrieval and alteration to suit the demands of the design. He equates experimenting with different formulas to playing the piano, which he did when younger. "If the music didn't sound right and I determined it wasn't me but the piano, I changed the piano. It's the same with getting the right mix for the batch. If it's not right, the glass will not be right."

It may be that glassmaking is too rushed now for Mark Peiser, that modern innovations such as automatic furnace controls are not necessarily an improvement over the time when a glassmaker put a hand on the outside of the fire wall to gauge the temperature.

The beauty and elegance of blowing glass is important to him, and he admits he was infuriated when Harvey Littleton said that technique is cheap. The striving to make art has overshadowed the importance of glass as a magical material, he believes, and he is wondering where to go from here. In fact he told me he had committed himself to something new, to going out to Colorado and try sculpting in marble and bronze.

"I haven't fallen out of love with glass, but all of a sudden I have a different relationship with it," he says. "I work my ass off to do the pieces I do. It takes time. I've heard the rallying cry that we must prove glass is a valid medium for art, but wouldn't it be better to address the question of what's the valid art to make from glass? If you want to do a piece called *Resignation,* you can't do that in glass. How can you do something called *Resignation* in glass? You have to fit the material to the subject."

James Thurber wrote, as only Thurber could, "He knows all about art, but he doesn't know what he likes," and that, I think, is what Peiser may feel about some of the glass people working today. For him, the greatest gift of glass as a material for art is its transparency, and he has used the gift well, as have many glass artists including one for whom Peiser holds high esteem: Jon Kuhn.

More than anything, a Kuhn work is a virtuoso performance in technology, but a technology that reaches across the calculated angles and refractive indexes to embrace a mesmerizing aesthetic value. There are things he does with cut glass and crystal that can skip light around like flat stones on water. He is a star in the studio glass art movement, as he should be.

Having purchased a Jon Kuhn art piece, Dr. Richard Basch and his wife, Barbara, were preparing for its arrival at their home in Florida (when shipped, valuable glass art comes carefully packaged, of course, but there is a moment of extreme apprehension when the last of the plastic and Styrofoam and balled newspapers and whatever else is stuffed in there is pulled out, hopefully without the heart-sinking discovery of breakage) by moving pieces of furniture and other art around to make way for the work. It was to be more enthroned than simply put down.

They are collectors of glass art, but more than that, they have allowed their lives to become like an accommodation for the glass that fills most of the rooms of their spacious house. "We live among the glass," said Dr. Basch. "It's with us, and it's very important."

Collecting glass is an old and honorable pastime for many people the world over, especially Americans who can mine the legacy of the late 1920s and all of the 1930s for the riches of Depression glass. That was the glass given away, along with money and other goods, on Bank Night at local movie houses, those not-so-grand theaters that forever smelled of Flit and the mustiness of worn carpet runners in the aisles. There were bowls and dishes and other dinner-ware given away not only at the movies, but at gas stations and furniture stores. It all came in many colors: anyone who has ever seen a pink or light green candy dish has probably seen a piece of Depression glass.

Certain pieces of Depression glass can fetch prices in the low four figures now, not because the glass is particularly attractive (it isn't), but because it is glass that sparkles with the history of drab times. Take the mug of blue glass with Shirley Temple's face on it. There's more to it than Ovaltine.

The type of glass collecting that the Basches do is not at all like that. They are in fact among a cadre of about 350 glass enthusiasts/collectors who have come on the scene since the inception of the studio glass movement in the 1960s, few if any of whom are interested one whit in a plate or tumbler unless, of course, it is incorporated into some grand sweep of glass as art. All of the artists represented in the Basch collection are alive, and to them that is important. "These artists deserve to be admired and to know they are admired," Dr. Basch said. "There's no better way to do that than by having their work in our home."

The collection they have assembled since 1989 is an eclectic one. They have three Jon Kuhn pieces now, along with at least one and often more works by many of the best known contemporary glass artists. They have William Morris, Stanislav Libenský, and Lino Tagliapietra, among others. All three are artists of high station in the studio movement, along with others whose names are taking the light as does the glass of their art: Dan Dailey, Joel Philip Myers, Howard Ben Tre, Ginny Ruffner, Dick Weiss, Jay Musler, Richard Marquis, Benjamin Moore, Sonja Blumdahl, Richard Royal, Joey Kirkpatrick, and Flora Mace.

And there are others, one especially. His pieces in the Basch collection are the ones most prized by the couple. The things he does carry the same high value throughout the world, for if you talk about the exuberance of the response to contemporary glass art, you talk about this man.

You talk about Dale Chihuly.

Chapter Thirteen

Dale Chihuly and a
Place Called Pilchuck

"I didn't sell a piece of glass until 1975."

Dale Chihuly was saying that, and smiling as he did, for in the years since the end of the dry spell, he has become one of the most financially successful artists of the twentieth century. Even as we spoke, he was preparing to do a glass sculpture for the headquarters building of a pizza company; his fee for the piece would be half a million dollars.

If there are strong pecuniary overtones to the Chihuly engine of glassmaking (and there are), it is not resented by others, for he, more than anyone, is responsible for the attention being given to the studio art glass movement. And besides, his enormous talent permits him some leeway for financial indulgence. He has had a one-man show at the Louvre, a rare achievement for an American artist, and had the first show in the new Seattle Art Museum. He has been scrolled and gowned, profiled and fussed over.

Chihuly, a man racing with creative energy, has a studio called the Boathouse, a place of forty thousand square feet, on a waterfront in Seattle, where he and assistants carry his designs to creation. There is no mistaking a Chihuly piece. It is usually oversize and swirling in opulence, a work of sensuous vitality to win the heart. The colors he uses seem drawn from the dazzle of carnival: translucent violet and golden celadon, cadmium yellow and cobalt, crimson and cranberry. Even the names of the series of his creations speak of grandness: "Seaforms," "Niijima Floats," "Venetians," "The Persian Installations."

Among the things that Chihuly makes are gigantic chandeliers of mouth-blown glass, and once, when chandeliers were in his thoughts, he concluded that hanging some of them over the canals of Venice would be a fine thing. Of course, this was not something that Dale Chihuly intended to keep to himself, for his instinct for self-promotion is acute. If it sounded like a trashy project—an Evel Kneivel-jumping-a-river-on-his-rocketbike sort of thing—that would be overcome by the fact that he has the artistic ability to back up his temerity to use the city where Titian was born as a place to hang his glass chandeliers.

There is this, too: Chihuly knew that by succeeding with the Venice installation, it would be a celebration of sorts for the emergence of decorative glass as a proven medium for art. He genuinely wants people to like glass.

Chihuly being Chihuly, it wouldn't do to simply blow the chandeliers in Seattle and ship them off to Venice. As he explained, "Why not go to five countries with great glass traditions and blow chandeliers with both American glassblowers and glassmasters from each of the countries—a collaboration of cultural techniques and talent that would be unique for each country? I would go with no preconceived ideas and take it day by day. The results would be impossible to predict, and each chandelier would be unlike any other. We would break down the cultural barriers and dispel the secrets that have so long restricted and insulated the great glass houses. In the end the chandeliers, made from thousands of individual parts representing the work of hundreds of glassblowers, would be sent to Venice."

And in the end, chandeliers—fifteen of them—were indeed sent to Venice, culminating a million-dollar project involving at least three hundred people. They were blown in Finland, Ireland, and Mexico. They were large pieces, up to fifteen feet in diameter and weighing half a ton, but it was the color, all the limes and pinks and others of the Popsicle shades Chihuly so loves, that give this installation its dynamic presence. If nothing else, these translucent confections could serve to remind Venice of its rich history with the making of ornate glass.

It was all so Chihulian. At Waterford, where some of the chandeliers were blown, he brought more excitement to that staid house of crystal-making than anything since the time in the 1780s when John Hill, the chief glassblower, quit after deciding he couldn't get along with the owner's wife.

He arrived there from Seattle with an entourage that included ten of the glassblowers from the home shop, his girlfriend/manager, a couple of television crews, and assorted others. Before long, with his wool cap on backward in the Irish way, and with the overall appearance and demeanor of a manic leprechaun, Chihuly was sketching designs on large sheets of paper and giving voice to the creative ideas that tumbled through his mind. Waterford employees stood around, watching, as a placid mourning dove might watch the helicopter dynamics of a hummingbird.

In Finland, too, there was fascination among the locals for the American who came to their village and hung beautiful pieces of richly colored glass from cables strung between the birch trees and put others in the river to float among the lily pads. Chihuly arrived in the village of Nuutajärvi, a place north of Helsinki with a long and proud history of glassmaking, with a crew of forty glassblowers and others from the boathouse. For seven days, they and the Finnish blowers made glass. They made chandeliers—one with gourd-shaped parts of a soft yellow, and another of cobalt blue.

Villagers went out one morning to find spears of red glass standing in teepee-shaped stacks along the riverbank. And there was Chihuly, standing on a small bridge that spans the river, dropping basketball-sized blown glass over the side; the spheres were greenish blue, about the size of a volleyball, and when the sun was up, they glistened and forged some sparks of life in that old and lovely river. Elsewhere, in a reedy place there suddenly arose coils of glass, each with a head like that of a cobra.

The people of Nuutajärvi, needless to say, had never seen anything quite like it. Later, Chihuly would write to the villagers, telling them, "It was a wild and wonderful time. I want to thank you all."

The installation in Venice coincided with the staging there of an international glass show called Venezia Aperto Vetro (Venice Open Glass), an event designed to showcase the works of contemporary artists. It would be out of character, of course, for a Venetian to show more than faintly discernible enthusiasm for modern glass art by an American. They count Venetian glass among the glories of their cultural heritage (that same pride of heritage has spun off progress-choking provincialism and inbreeding in the making of glass in Venice), and day in and day out, they live with the treasured art that makes Venice so special.

The city itself is a piece of decorative art, and a dozen or so glass chandeliers hanging over the canals could not make much difference one way or another. Christo could wrap the city in wet manicotti and the reaction would be the same: a shrug, a sigh, dismissal as folly. What does arouse the interest of an Italian born and raised in Venice is women (for a man), clothes (for a woman), and the puzzling insistence of foreign tourists on sloshing around in water up to their knees when the city is flooded (for both).

Dale Chihuly is in his mid-fifties. He is a short man with a large upper body, a soft, very round beach ball of a face, and a great shock of curly hair. He has worn a patch over his left eye since 1976, when he was in a near-fatal automobile crash in England. With the loss of one eye he also lost his depth perception, and without that, it is difficult to blow glass.

And that's the unique thing about Dale Chihuly, generally conceded to be the most successful glass artist in the world today: He doesn't blow glass.

Among other glass artists and experts such as collectors and gallery owners, the admiration for Chihuly and celebration of his artistic talents are not diminished by the fact that others give life to his ideas. Rather, this is what is heard: "He is the maestro, the others the orchestra," and, from one of those who blow glass at the Boathouse, "He is the wind in our sails."

In a way, Chihuly works in the tradition of the masters of the Renaissance and their teams of apprentices. Many of Tiffany's pieces came from the hands of his workers. It is doubtful that Chihuly could have achieved the success he has had he been working alone through the years. He has said many times that working alone is simply not for him. Much of his energy and stimulation comes from the presence of people. Also, with the size of the pieces he makes, the work must necessarily be an ensemble production. There is a staff of some seventy-five people—glassblowers and others—at the Boathouse, and the atmosphere there is an amalgam of corporate headquarters and summer camp. There are offices and secretaries, and some midlevel people with voice mail.

One thing that shouldn't be missed in the Boathouse is

the $500,000 lap pool. It is fifty feet long and has a large grouping of lighted and brilliantly colored Chihuly art glass (most of it shell-shaped) on the bottom. It is all covered by a sheet of safety glass, so swimmers can move right across the art. Away from the pool, there is a table to be seen, an eighty-seven-foot-long table carved from a single Douglas fir tree. It is five feet wide and rests on eight steel supports. Seeing this table, you have to think that the Council of Trent could have convened around it.

Chihuly is in a large room at the entrance to the Boat-house. The wooden floor is splattered with paint, as he is himself, with his trousers and shoes stained in such a way as to be almost art in themselves. On the floor are large sheets of paper on which the artist has sketched designs for chandeliers and other works. A television cameraman and a producer are in his footsteps, recording his movements and words. He is talking with an assistant about a sketch, about a choice of colors, about a new idea he has. He slashes at paper with a stick of charcoal, and the strokes reflect the bold creativity that's taking place. Dale Chihuly is indeed creating, and he is doing it with a degree of crisp definition rarely seen among glass artists. Many of the designs will be discarded, but some will go out to the blowers, who will gather molten glass and give form to the man's ideas. And there will be loud music playing when this is done, and Chihuly will be there to advise, to joke, and to cheer when the flawless pieces are carried off to the annealing lehr.

"There are good points and bad points about a busy life like this," he said, pausing to sit and drink coffee. "The thing is, I've never lost my interest in glass, and the limits to which the material can be pushed for art are still far off." If his pieces for the most part are large now, they will proba-bly only get larger in the future. He once did a set for an opera, Debussy's *Pelléas et Mélisande,* in transparent fabrics, and it is easy to envision him doing this again, only with glass as the material.

As to whether his work is craft or art, he says only, "The best of any craft is art."

A year after graduating from the University of Washington in 1966, Chihuly went to the University of Wisconsin to study glassblowing under Harvey Littleton, and by 1969, with advanced degrees from Wisconsin and the Rhode Island School of Design, he was in Europe, on pilgrimage to such masters as Stanislav Libenský and Jaroslava Brychtová in Czechoslovakia, and Erwin Eisch in Germany. By that time, too, he had started to receive the first of many grants and teaching fellowships. One of the fellowships, a Fulbright, allowed Chihuly to apprentice as a glassblower at the Venini factory in Murano, the first American ever to study there.

It has continued with hardly a break: new challenges taken on successfully, honors piling up, works prized by museums and private and corporate collectors, growing fame.

In 1980, Chihuly began his now famous "Seaforms" series. With the appearances of these pieces, the studio art glass movement moved onto high ground. Of all approaches to the creation of art, maybe none is so right for spontaneity as glass blowing. Putting human breath into a glob of molten glass can cause wondrous things to happen if there is a compromise between the artist's control and the tendency of molten glass to form freely when swirled. With Chihuly's "Seaforms," all of the best of spontaneity and control came together in a way never seen before.

The "Seaforms" are lavish with color—scarlet pieces lipped with yellow, coral pink with black lip wraps, and so forth. Many of the pieces have the gaping mouth pose of certain sea creatures, and others the fragile floatings of lilies and their pads. The glass is sometimes rippled and pleated, and always seductively iridescent. Chihuly has washes of the sea in his background, including a stint as a fisherman in Alaska, and that no doubt served him well in making these works. In the book *Seaforms,* Sylvia Earle, one of the leading oceanographers of the twentieth century, writes, "Whatever else these wondrous glass objects are as reflections of skill, passion, teamwork and sheer genius, they are also tributes— a celebration of the sea."

It is surprising, I think, that so many years have passed since the introduction of "Seaforms," and Dale Chihuly has yet to go off to the Great Barrier Reef and exhibit his glass there, underwater. All of the color and all of the indissoluble strength and beauty of the "Seaforms" should do those sad depths, where the coral lies ailing, a world of good.

There was another sea-oriented series, "Niijima Floats," and the idea for that was drawn from Chihuly's childhood memories of the Pacific near his Washington State home, where the floats for Japanese fishing nets washed up on-shore. The massive (some three feet in diameter) works are not so colorful as "Seaforms," but they will never be mistaken for oil drums. As much as anything, they are a solid testament to the strength and skill of the glassblower, for working hot glass of that size is a major challenge.

Of all his works, none have been so bold as those of the "Venetians" series. Venice has been an inspiration to Chihuly as long as he has been working with glass. He has a romance with the city, and the "Venetians" pieces themselves are full of romance. They are vessels of mostly ordinary form, but any simplicity is smothered in the great wrappings of decoration on each piece. There are flowers and fish, and golden cherubs, and wonderful ribbons tied as loosely as a bow on a poodle's ear. These are pieces to spur the shy to undertake adventures of the heart.

There is a steady movement of people through the big

room of the Boathouse, where Chihuly is running his art business as I suspect Bernini and Michelangelo ran theirs. "Do you have the layout for the architectural project?" he asks a young man. There is another man who is wearing a jacket, necktie, and shorts. He may be a television producer. Chihuly warns me not to drink too much of the very good coffee available in Seattle; it kills sleep in its tracks. He is dressed in a dark silk shirt and jeans the color of an egg yolk.

Dale Chihuly is the perfect point man for the studio glass art movement. He has delivered the cachet of respectability long denied the material by museums and galleries. He has thrown his work in the face of the world and dared the world to naysay its artistry.

Chihuly counts among his admirers Hollywood celebrities and business types with more money than the treasury of Malta. In Washington, Hillary Clinton has paid honor to the artist by putting one of his works in a prominent location in the White House. He has honorary doctorates, awards upon awards, and has even been honored by the University of North Carolina, Wilmington, as a "National Living Treasure."

It's not enough. Chihuly keeps running, living his breathless life as if it will all end tomorrow and he'll have to return to Tacoma and work nights in a 7-Eleven. In 1992 he told me, "I'm barely breaking even." Five years later he said things were better, although his expenses remained high. The Venice chandelier adventure, he said, cost him about $1.5 million. Artists, of course, are expected to be ambivalent about money matters, but when success cascades down on a person, as it has on Dale Chihuly, the entrepreneurial spirit is likely to be stirred to life.

When last heard from, Chihuly was preparing to open a bookstore in Seattle, with a restaurant in it featuring Sicilian food prepared by Sicilian women—a place of a hundred thousand square feet, with the artist's collection of more than five hundred accordions hanging from the ceiling.

Seattle wears its resident celebrities well, and they in turn have deep-rooted attachment to the city. Chihuly is among them, as is Howard Schultz, CEO of Starbucks Corp., and Bill Gates, Microsoft's billionaire chief with a cowlick. The latter two gave the Seattle area a corporate aura (and an enviable tax base), while Chihuly gave it bragging rights to the validation of glass as a medium for contemporary art.

Chihuly did that when he established a school for glass-working called Pilchuck.

Although it is about fifty miles north of Seattle, Pilchuck lies deep within the embrace of the city's affair with glass art. Indeed, without Pilchuck, Seattle would likely not be the center of the studio movement. There have been other schools for the teaching of the craft, but none so exclusively devoted to glass—none with the prestige to attract the titans of glass art to serve as instructors—as Pilchuck. Today, it is the leading learning center in the world for the creation of art with glass.

In 1971, charged with the excitement of having taught in the glass program at the Haystack Mountain School in Maine, Chihuly set out to establish a school in his native Northwest. He found a site of fifty acres in the foothills of the Cascade Mountains, a green and timbered place with views that carry down to the waters of Puget Sound. He enlisted the financial support of John and Anne Gould Hauberg, of Weyerhauser lumber money, and with funds in hand, construction began.

Twenty-five years later, Pilchuck, having taken its name from a creek that flows across the scenic acreage, is a thriving school with applications for admission running far in excess of openings. Ann Jacobson, registrar at Pilchuck, cited some of the foreign countries represented in an incoming class: China, Australia, South Africa, Turkey, and most of the European countries. The majority of those who attend are professional glassworkers.

Pilchuck is open for classes only in the summer, and just to be there then, on those high, cool grounds soaked with the smell of fir, is worth the tuition cost itself. It is near the end of a session at which some fifty students have been in attendance, and there is a sense of muted jubilation among them. They have sat here at the knees of the masters, blowing glass with them, making art with them, stuffing themselves at that feast of talent.

One student says to another, "I think I always knew it, but being here has convinced me that glass is so inherently beautiful that you can make a dog turd out of it and it would look good."

Jan Mares is a Czech, an engraver of glass, and a former student in Prague of the renowned glass artist Stanislav Libenský. He is at Pilchuck most summers as an instructor, and even on his own time he can be found at the diamond and copper wheels, cutting and polishing, enfolding bold images in layers of crystal and hoping, always hoping. . . .

"It is your piece, and you are almost finished with it," Mares said. "Then you do a little final polishing and, you know, make the mistake. You have been with the piece for a month, two months, and then in only a second, at the very last, it can break. But, you know, I think that glass is a special language for artists. Now glass can be like a painting, a sculpture, but not many years ago, people thought that glass is only something to put on the table, just like a vase, a plate. Just pieces for using.

"But now it is big art. People like Chihuly, they start to make glass like art, you know. Forget it for the vases for using flowers, and just make some objects and show what is possible with glass."

The people at Pilchuck are truly rapturous over glass, as Dale Chihuly, I'm sure, knew they would be. In the very early years of the school, those who enrolled had a narrowly limited knowledge of glass, simply because masters at that time—just twenty to twenty-five years ago—were still keeping the working of glass secret. Few dared breach the unwritten covenant of silence.

Lino Tagliapietra was one who did.

Most anywhere where there are glass art people in large numbers—Seattle, Murano, the mountains of North Carolina—the name Lino Tagliapietra is often heard. Many consider him to be the best glassblower in the world. He is Italian, from Murano, and when he first came to Pilchuck in 1975, he came as one who had turned his back on the secrets so closely held on the island of his birth. He revealed the techniques for making Venetian glass, and his fellow masters in Italy ostracized him.

"The people on Murano were very angry with me for coming to the United States to teach," he said to me as we walked about the grounds at Pilchuck. "They talked bad things about me. Considered me a traitor. They wanted to keep the secret. But just as the secret came, the secret went."

It has taken more than twenty years, but Lino Tagliapietra thinks he has been taken back into the good graces of his fellow glassworkers on Murano. "My grandson—he is nine—wants to be a glassblower," he said. "I will encourage him. I'll probably build him a small furnace in Murano and teach him there."

Moving down the front half of his sixties now, he is a powerful man with a body shaped by a lifetime of exertion in the hot shop. He teaches advanced glassblowing at Pilchuck, and if just one of his students can attune himself or herself to the rhythm of Tagliapietra's technique, the rewards will be ample. Without question, he is the star at

Pilchuck. "He has almost single-handedly revolutionized glass-blowing techniques, allowing students to learn like they never were able to do before," an admiring female student said.

A class in beginning glassblowing has come to an end, but those in attendance remain seated in the open-sided shed where the furnaces are located. It is as if they had attended the preliminary match and were anxiously awaiting the main bout. Tagliapietra is due to appear. Among his advanced students who have positioned themselves close to where he will be working is Nadège Desgenetez, a French woman from Normandy who is a champion of the emergence of women as a strong force in the future of mouth-blown art glass.

And Lino Tagliapietra recognizes that. "Women are better students than men," he said. "I think in the next millennium women will be the greatest glass blowers of the world. You know, glassblowing in many ways is like ballet. The glass is the music. Women have the proper attitude for great glassblowing."

He began at the age of eleven, and by twenty-one he had become a maestro. "My uncle was a master, and my cousin a master," he said, "but no one in my immediate family was blowing glass." While others dream of apprenticing at Venini in Murano, Tagliapietra can say he was a master blower there, a gaffer at one of the most respected furnaces in the world. It was because he was visionary enough to see an expanded role for women in the profession that he evoked the scorn of many on the island.

Unlike most gaffers, he does not wear dark glasses when making gathers of molten glass through the glory hole of the furnace. It isn't that he is any less protective of his eyes, but rather that his exchange with the fire leaves no room for mishap. He started to work on a piece and he was with it for forty-five minutes, going back and forth to the furnace to reheat, rolling the gather over the flat surface of a steel table, blowing and shaping until it was long and tapered, then reheating and reshaping until it was squat and bulbous. Someone brought him a cup of coffee, setting it on the floor by the bench where he sat when blowing (in Germany and the Republic of Czechoslovakia, glassblowers stand when they use the pipe). He finally had the shape he wanted and so he began adding the decorations, drawing glass down into fine threads.

Actually, through all of that, he remained somewhat of a spectral figure before the students, his movements ghostly perfection as he worked the glass into what became at the end, a lovely vase with all the delicacy and finery of Alençon lace.

The woman from Normandy assisted him, and when it's over and the students applauded, she backed away from his side, and back some more until she was off center stage. And then she too applauded.

Later, Tagliapietra demonstrated some of the fine points of glassblowing for me: the use of the jacks, or *pucellas,* to shape the glass during the blowing process; pressing the free end of the gather flat to form the bottom of the piece; reheating; drawing glass into thin threads for use as decoration—doing all the things that make the shaping of molten glass a magical thing.

Of course, I itched to try my hand at blowing glass under his supervision, but I couldn't bring myself to ask about that. It would be like asking Cal Ripken to play catch or Pete Sampras to hit a few.

Because he is renowned as a blower of glass, Lino Tagliapietra's talent as a designer and artist is sometimes overlooked. "I don't know if I am an artist or not an artist," he said. "I do know that now when I blow for a factory in Murano, they put my name on the piece, not the factory owner's name. Now, yes, my name, but only after fifteen, twenty years."

And so they come each summer to the school that Chihuly founded, hoping, of course, that they can do what he has done—people like Brady Steward of New Orleans, thirty-four years old and working on a master's degree in glass sculpture at Tulane University. "I took some art courses in college and one of my teachers told me I should get my ass in glass," he said. "For the first four years I worked with glass, I was mesmerized by it. But then I got control, and now the glass works for me."

Pilchuck, now a striking cluster of buildings made mostly of glass and cedar, is the hub for the greatest concentration of glassblowers outside of Murano. Surprisingly, there is a feeling of union among these people; seldom does one speak harshly about another's work. They even wish Dale Chihuly well in his relentless quest for bold exposure. Much of that, no doubt, has to do with the youth of the studio glass art movement. They simply haven't had time enough to raise up a good head of artistic spite. In fact, it was being said back in 1993 that the goodwill had started to erode, but there's not much indication of that.

"Glass people tend to stay together," said Seattle art gallery owner William Travers. Yes, and the circle is widening. Because of Harvey Littleton, Dominick Labino, Dale Chihuly, and Pilchuck, the United States has become the hub of glass art making worldwide. Lino Tagliapietra concedes that, but when he talks about it, a note of sadness slips into his voice, a lament for the dimming of the glory that was Venetian glass.

"We have in Murano in the past the incredible defense of the secret of making glass. But now I think it is different. The big problem there today is that they don't care about a school for teaching glass. They don't ever want a school for the glass because they think the factory is enough for everything. Maybe it could be that way in the past. For me, I can't believe it. The school is always necessary. There is no vision in Murano."

And yet it is Venetian-style glassblowing that Tagliapietra teaches at Pilchuck. It is Venice that has a spiritual hold on Dale Chihuly. And where will Chihuly's work go in the future? Some fear that his instinct for the oversized and luscious will turn against him. He likes to say he can only do his work if he continues to expand the limits of glass as a medium for art.

That is heard so often: pushing the boundaries of glass. Sometimes it seems there can be a danger in doing that. If the technique overwhelms all else, is it then art? Have artists so soon exhausted use of glass with unpushed boundaries?

Certainly, it will be interesting to see where Chihuly moves next. Perhaps he will renew an interest in architectural glass, such as that now being made so successfully by a man named James Carpenter, with whom he collaborated to make glass art many years ago.

Chapter Fourteen

The Glass Art of the Architect

There are artists working with glass who become like hunters, setting traps for sunlight in that interstitial space between the outside and the inside. For some, that is all that matters.

James Carpenter, who thinks of glass as a dynamic tool with which to control light, is responsible for a spectacular mural of decorative architectural glass. The piece gathers in not only sunlight, but all of its nuances at all times of the day and in all seasons of the year. It can be seen on the Columbus Avenue side of Lincoln Square, a commercial and residential complex in New York City. At a hundred feet wide and forty-six feet tall, it shouldn't be hard to miss, but sometimes sunlight falls dimly in those urban canyons, and then the mural shuts down.

"Depending on what you want glass to do for you, it will respond in the appropriate way," Carpenter said. Obviously, he wanted the work to choreograph sunlight in an elegant review, as only glass can. The base of the mural is a panel of flat, semireflective glass that has been laminated and tempered; it is secured flat against the wall of the building. Ar-

233

ranged perpendicular to the base are 216 narrow blades made of laminated dichroic glass, or glass that has been coated in such a way as to allow designated wavelengths within the spectrum of light to be transmitted or reflected. There are formulated sequences for applying the metal oxide coatings, and if that procedure is successful, the results can be wondrous celebrations of light and color.

There have always been splashes of sumptuousness among the architectural decorative glasses, but some of these new-age works have moved beyond that to become defining pieces of art. Many of them could just as well be in museums rather than the lobbies of office buildings or the domes of malls and civic centers.

Technology and artistry play equally important roles in the making of modern architectural glass, and with his overriding concern being for the display of light, and for the use of glass in that regard, James Carpenter's talent spans the divide between the two. He knows how to blow glass and how to make magic of its reflective and transmissive qualities—the magic that can allow a skyscraper with cladding of highly reflective glass to vanish in the sky, leaving behind nothing but shimmering images of the neighborhood.

Carpenter, who once worked in collaboration with Dale Chihuly to produce a few artworks in glass, squeezes old technology for new applications. He envisions glass not only as decoration, or window and cladding material for buildings, but as a structural force within itself. Rather than structural frames, he wants, say, the curtain wall of a building to be supported by the metal and other strong laminates within the glass. Like most of his work, it involves the union of the functional with the aesthetic, and a sweet marriage it is.

He was called to the Christian Theological Seminary in Indianapolis (not a divine summons) to design windows for the chapel. They didn't want stained glass, but they did want light and color, so he designed a grid of glass with horizontal bands made of dichroic glass. That would be the window.

But not just that. Carpenter had drawn on computer-aided technology to calculate the schedule for shadow and light throughout most of the day. Balancing that with the ability of the dichroic glass to selectively transmit and reflect, Carpenter made his installations, and from that time on, light and color have played on the walls of the chapel like a cyclorama of spirituality.

A glass scientist once described for me what he thought would be the perfect glass. "It would be a reflector some-times, a transmitter sometimes," he said. "You would want it to be clear, then you'd want it to absorb the infrared and pass the visible. You can throw in the equations and calculations, everything you want, but we don't have a glass like that yet."

James Carpenter's coated dichroic glass comes close, as do some other forms of the material being used today for architectural purposes. It is by the grace of recent advances in glass technology that renowned architect I. M. Pei's de-sign for the pyramid entrance to the Louvre in Paris could become a reality. That same technology has permitted re-moval of the unsightly wire glass dome and three nets used for many years to protect the treasured glass dome on the Milan Borsa.

The possibilities of modern architectural glasses are enough to embolden the least venturesome in design. The worthy Lloyd's of London, for example, may stir thoughts of cutting-edge risk coverage, but hardly of innovative archi-tecture. Glass is helping change that. The insurer's latest headquarters building, of four built in the twentieth century, has the architectural boldness flaunted by the Pompidou Center in Paris, and that may be because the hands of one man, Richard Rogers, guided in both designs. His firm,

Richard Rogers Partnership of London, imprints the world of design with its high-tech stamp.

Mike Davies, one of the Rogers partners, is an expert on the use of glass in architecture, and he is confident that the material has much to offer before reaching its full potential. "Glass is one of the longest-lived materials of low technology," he said. "But with the advancements being made in microchemical technology, buildings will become revolutionized by the use of glass. It is the great building material of the future. Already, the glass used in the new Lloyd's building has been superseded by new glass materials."

It is said of Lloyd's that the company insures 75 percent of everything that moves, and to be in the main trading hall of the building at eleven o'clock on a workday morning tends to confirm that. At that time each day, there are as many as seven thousand persons there doing whatever underwriters do. Occasionally, the firm's famed Lutine Bell, positioned in the room on an imposing wooden rostrum, will be rung to signify news of importance to the insurers. Two rings signify good news, one for bad. (I have a friend whose sailboat, insured by Lloyd's, went aground on a reef and was damaged beyond repair. Now when he enters a room, there is a freeze of hushed awe, and this whispered in admiration: "They rang the bell for him at Lloyd's of London.")

The core of the building is a 167-foot-high atrium clad in glass that is double-glazed with a gap in the middle. Warm air is directed into the cavity between the two layers to maintain a constant temperature on the inner wall. The energy saved with the use of such glass is substantial—as much as 60 percent in some cases.

But it's something much more visible than a glazed sandwich of warm air that causes Londoners and others who venture along Leadenhall Street to stop and stare. What they see is a glass curtain in which thousands of prisms were rolled to forge diamonds out of the sunlight and add sparkle to that towering house of indemnities.

And it will get better than that, according to Mike Davies, for the time is coming, he believes, when a person can look up at a spectrum-washed glass skin "whose surface is a map of its instantaneous performance, stealing energy from the air with an iridescent shrug, rippling its photogrids as a cloud runs across the sun; a wall which, as the night chill falls, [turns] white on its north face and blue on its south . . . clears a view-patch for the lovers on the south side of level twenty-two and . . . turns twelve percent silver just after dawn."

That may be more than we really need from glass, but the material does tend to arouse the visionary spirit of architects.

There have been many advocates for the use of glass in architecture, but none quite like the late Paul Scheerbart, a German writer of science fiction. He was possessed of excessive zeal for the material, even to the point of ascribing therapeutic powers to it. In an environment of glass structures, he believed, humans would become better people—less warlike, free of petty jealousies, smiling and happy, and, who knows, maybe even more giving of alms. That aside, Scheerbart, who died in 1915, did indeed see many qualities in glass before scientists got around to uncovering them. It was under his influence that another German, Bruno Taut, directed the organization of fourteen German intellectuals into a group that called itself the Crystal Chain, and during the latter months of 1920 and all of 1921 they pushed their fantasies of works in glass to the limits, circulating a letter among themselves with their drawings and thoughts set down in an extraordinary record of devotion to architectural glass.

It was all so wonderfully visionary: glass bubbles in the sky launched from airplanes, great cathedrals all of glass,

glass houses in which there would be no walls on which to hang a framed picture. And they wanted to publish a book that would reduce all religions, all wars, all lusts and sinful recreation to nothingness. Among other things, this great book would call for the onanist to give up his phallus.

Understand, the Crystal Chain had genius and great architectural talent among its members. In all of the excesses there were solid cores of valuable contributions to the development of modern glass architecture. Look today on the glass-clad towers in New York—look especially at the Seagram Building—and know that the brilliant madness of the Crystal Chain has been committed there.

It was in that year of the Crystal Chain, 1921, that another German architect, Ludwig Mies van der Rohe, was having thoughts about an office building with walls of glass curtain. After serving as director of the Bauhaus School of Design, standard-bearer for modernism in architecture in the early 1930s, he came to America and built structures in Chicago and New York that gave new meaning to the use of glass in architecture. With another architect, Philip Johnson, Mies designed Manhattan's Seagram Building, the mid-1950s forerunner of an epidemic use of glass as cladding for towers. Many came to deplore what they called the "glass boxes." With the invention of the float process for making flat glass, many of the restraints on architectural use of the material vanished, and so miles and miles of it were raised as cladding, wrapped around steel framing, made to reflect or just hang there in milky opalescence or light-transmitting submissiveness, and, yes, cities were disfigured by it all.

But that was then, and through the passage of time and advances in technology and architectural skills, cladding glass has redeemed its honor.

If nothing else, the glass buildings of that time permitted dwellers in high-rise apartments and office suites to become accustomed to an environment in which the outside and inside are like participants in soft-core porn: they are positioned together, but without penetration. Glass is there to prevent that, but by its magical power of reflection, it allows a crossing-over of spectral images—the ghosts of party-goers and late workers aloft in the blackness on the other side of the light.

Of all the concepts in architecture, is there one more brilliant than that which allows tripping to the edge of the great maw of space without fear of falling in? We need not fear because the glass will hold us back. And why should we fear, since, in truth, it is we who are intruding on space, and not the other way around.

Here is Richard Sennett, a novelist, lecturing on glass at the Massachusetts Institute of Technology: "For the first time in a building by Mies I felt comfortable leaning back against the glass. I don't want to make too much of this moment, only that it gave me an intimation of what the word *modern* might truly and positively imply. It gave me a sense of the inherent ambiguity of glass; more than a metaphor, glass is a field on which the exchange between inner and outer occurs, a field reflecting the violation of space, but also enclosing and protecting."

Glass, of course, continues to be in heavy use as cladding in corporate architecture. Two towers recently risen in New York attest to that, and also to the fact that cladding glass is being used in even more innovative ways. The Vuitton Tower on East Fifty-seventh Street has been given a wall of glass that follows the contours of a setback at the tenth story, and then it goes up with a second layer of glass that slants off to reveal a third. Some of the glass is clear, some opalescent white, and at night it is all bathed in softly colored light from hidden neon tubes. It is like a huge glassy flower. The other building, the Austrian Cultural Institute on East Fifty-second Street, is not so horticultural. Rather, it has a front of glass panels that hang tilted in such a way as to seem almost threatening. One almost expects to find a basket on the sidewalk below—something to catch the heads. Sinister or not, it is an elegant structure.

Of all those who have put faith in glass as the skin for their towering monuments to self and economic enterprise, few if any can equal Sir Norman Foster of England and his proposed 1,265-foot-high London Millennium Tower. The use of glass in that high-rise would showcase transparency in unobtrusive clarity. There will be no massive steel framing or elevator shafts to prevent open views from the front through to and out the rear. But for the narrow steel supports around the perimeter, the walls will be almost all glass—double-glazed glass with space enough in between for air to circulate and assist in temperature control. As for the elevator shafts, they will be, in the words of one of the design engineers, "no more than holes in the floor."

Plans were to have the building completed by the start of the new century, but that goal is not likely to be met. By late 1997, permission to start construction had not yet been given by the city. If and when the London Millennium Tower does rise, it will show glass in its boldest intrusion on space yet. There are those who fear it will also stand as a tempting target for the bombs of terrorists (Timothy McVeigh was quoted as having said that the glass front of the Oklahoma City federal building was a factor in his choice of a government structure to bomb).

In Pittsburgh, there is a glass house of commerce in the guise of a cathedral. And glass it should be, since it is the headquarters complex of PPG Industries. The six buildings of the complex are sheathed in close to a million square feet of double-paned glass. And all of that glass has been treated with reflective silver coating. For shimmering exuberance on a sunny day, this is an assembly of glass to be reckoned with. Certainly, there is nothing else in Pittsburgh quite so grand.

Something occurred on May 1, 1851 in London that caught the world's attention. In his "May Day Ode," Alfred, Lord Tennyson, wrote of the event: "A blazing arch of lucid glass / Leaps like a fountain from the grass / To meet the sun!" Charles Dickens wrote of it, too, along with others down through the years, until now, the 1851 opening of the Crystal Palace in Hyde Park has become a well-recorded milestone in the history of architectural glass.

The Crystal Palace was a sharp departure from architecture of the Victorian era; more than that, it was the first building of what could be deemed modernist architecture. It was all glass and iron, a massive structure designed by Joseph Paxton. Using his experience as head gardener to the duke of Devonshire and as one who had experimented with different ways to build a greenhouse, Paxton had the structure of approximately 1,850 feet long raised in just four months. The amount of sheet glass used was an inventory-sapping 900,000 square feet, and it was all mouth-blown glass that, as cylinders, had to be split and ironed flat.

Millions came to visit the Crystal Palace and the products on exhibit there. And they marveled as much at this use of glass as they did at leather products from Spain. Following the exhibition, the Crystal Palace was dismantled and reconstructed on Sydenham Hill in the south of London, where for eighty-two years it served Londoners and others as an arboretum, an exhibit and concert hall, and a place to dine. Then, on the last day in November in 1936, a spectacular fire completely destroyed the Crystal Palace. The skies over London were lit that night as this temple to Victorian pride came apart. The iron girders were left twisted and folded. And all that glass, melting and flowing in streams, only to cool and re-form and take up station in an ungodly landscape.

The legacy of Paxton's building has less to do with aesthetics than with construction techniques. It brought together the use of glass, iron, and wood for the first time in such a way that it all locked together—the two hundred miles of glazing bars, thousands of tons of cast iron, a forest of timber—and stood in such a way as to be truly a palace of glass. Before it was opened to the public, the Crystal Palace was tested for sturdiness and stability by having platoons of soldiers double-timing back and forth across critical areas in the building. It withstood it all, and it is on that firstborn of modern glass that architects and engineers have been building in technique and design for the past century and a half.

I went one summer to London's famed Royal Botanic Gardens at Kew, and that was when I first saw the Palm House, a glass structure as pleasing to the senses as glass structures can be. It should be said here that I have deep regard for botanical buildings, especially the old conservatories, those great beached whales of glass. And of them, the Palm House, to my thinking, stands in honor as one of the most remarkable achievements in glass construction of all time.

With its sixteen thousand individual panes of glass, the greenhouse takes the strike of light in blinding bursts at times. At other times, the light lies on the building as softly as a candle's flame. Snow and rain and the leaves of autumn fall there, too. And so it is a fine place in which to retreat, even to one of the rooms of stifling heat where plants of invasive nature grow, and look up at the panes of the ceiling and watch glass in service as a sketch pad for nature.

So many of the conservatories in England came about because of a need to house the shiploads of plants and other collectibles brought back from far and exotic places by members of expeditions during the seafaring days of the empire. First they built the greenhouses called orangeries, and one of the great ones is at Kew. "This is called the Orangery," Peter Riddington, an architect formerly employed at Kew, said as we stood in front of a modest glass-fronted building. "The glass here has been replaced, as has much of the glass in the conservatories here at Kew. Orangeries were built because as citrus plants were brought in from the Mediterranean—oranges, lemons, limes, and all those things—there was a need to shelter them in the winter, because otherwise you had to shroud the plants in sacking or whatever to keep the frost off. By the late eighteenth century, the more traditional greenhouses, as we know them, had appeared."

Designed by Sir William Chambers, a legendary figure in the history of British architecture, the Kew Orangery was constructed for the dowager Princess Augusta. It is a long, elegant building, but the use of glass was limited in its construction, and therefore so is its utility as a place in which to grow plants. Not far away is a weeping beech tree that had been damaged, most likely by the gale winds that ripped through Kew in October 1967, causing extensive damage to trees and other in-ground plantings, but little if any to the glass greenhouses. I suspect that the beech has been there for a century or more.

Kew Gardens is a joyous place in which to walk. There are three hundred acres there, and scattered throughout them are, in addition to the Orangery and Palm House, the reconstructed nineteenth-century Temperate House, now a large and magnificent hipped-roof pavilion with two wings, all in glass, and six other greenhouses. As in most public gardens of long and proud history, the plantings seem to be healthy almost to the point of rudeness. There are no aphids here (that I can see) to suck the sweet juices from the roses that bend stems with their weight. In another greenhouse, a

Chilean wine palm has reached a height of fifty-eight feet and a weight of fifty-four tons. That in itself may not trigger spasms of amazement, but consider this: It was grown at Kew *from seed.*

"One of the problems at Kew is that they don't want plants to just grow all right," Peter Riddington said. "They want them to grow brilliantly. Water is always being thrown everywhere."

The Palm House resembles an overturned boat, although it is sometimes called a "Victorian jelly mold." However it is seen, it deserves its recognition as the finest Victorian greenhouse ever built. Designed by Decimus Burton and engineered by Richard Turner, the curvilinear structure of glass and cast iron has a timeless charm. The builders of the Palm House benefited from new improvements in the quality of sheet glass. For many years in England, crown glass was used in greenhouses, a glass made by blowing a large open-ended vessel of the substance, and then spinning it until it opens into a large disk. It was a difficult way to make glass, and the waste involved in cutting squares from a round piece of glass was substantial. But crown glass had a pleasing luster to it, and that was cherished.

The Palm House was first restored between 1955 and 1957, and again, at a cost of more than $20 million, between 1985 and 1988. The replacement glass gleams now, and it has been set in silicon rubber mastic rather than putty. Special attention has been paid to giving renewed life to the interior of the house, for in a greenhouse, with its cloning of rain forest conditions, most damage to the structure occurs inside. The rust alone is enough to dismay a preservationist.

They did it all, though. They returned the Palm House to all its glory, and maybe more. Not only is it tickety-boo to the patrons of Kew, but in the world of architectural glass in general, it is a celebrity of infinite stardom.

In America, an admirer of glass conservatories would have to go to the Bronx in New York to find one comparable to the Palm House. There, across from Fordham University, is the New York Botanical Garden, and that great museum of plants and trees, a refuge sought each year by half a million visitors, has provided inestimable benefits to horticulture in America. There are people living in the Bronx who have been escaping to the Botanical Garden for relief and renewal all their lives. Sometimes they come to spend a few hours in the warm, glassed-in old Victorian conservatory, and sitting there, close to where orchids bloom beside waterfalls, they dream of having gone to Florida for the winter.

The conservatory there is named for Enid A. Haupt; it reopened in early May of 1997 following a four-year restoration project that cost $25 million. Nothing in the Bronx has looked as good as this for many years, not even Yankee Stadium, and it would be a pity to pass this life without seeing it.

It is a winter garden under seventeen thousand panes of glass, fronted by a ninety-foot-high domed rotunda. At first, the glass was clear, but it now has been tinted a pale green to soften the sunlight. There are about ten different display areas in the conservatory, and between them, the horticultural offering is vast.

Home-sited glass conservatories are rarely seen in these times. They are still being made, but the cost for a good one is high. Frames wrapped with plastic serve most gardeners now. And it may be that the last great conservatory of botanical-garden quality has been erected. But in spite of all of that, the compatibility between glass and gardening will remain invulnerable. There will always be an African violet by the kitchen window.

Indeed, there is an artist in New Jersey who transforms the relationship of glass with nature into exquisite works of glass art. Paul Stankard is a master of the centuries-old technique called flameworking, in which there is an intimate working with rods of glass using a gas-oxygen torch for making the glass soft, and small, handheld tools, such as tweezers, for shaping. The color is predetermined and available in the various shades of the rods.

Flameworking was once regarded as a craft to be practiced at carnivals or on the street, something by which to fabricate a two-dollar replica of a turtle. Stankard has destroyed that trivialization with his arrangements of nature encased in clear glass. Some call them paperweights, but to use one of his works to hold receipts and bills of lading in place would be outrageous. His pieces, seldom more than seven inches high, are craft risen to high art. And high price: do not expect to pick up a Stankard work—he calls them "botanicals"—for much less than $20,000.

A bromeliad in a conservatory will offer its spiky flower to the glass-passed sunlight in a cameo of nature in perfection. Stankard's botanicals come close to that, in that they showcase the world of plants and insects and birds in exquisite detail—the yellowing on a leaf, the gentle presence of a damselfly on a water lily. Those who come upon his art for the first time almost always assume that the pieces are merely glass in which real flowers and bugs have been preserved. But it's not that. If there is even something so small as a bee in a Stankard paperweight, it has been fabricated, down to flameworking glass for every one of the tiny, fuzzy stomach hairs. Many of the things he does are precise botanical studies, including root structures and environments.

"I'm into the spirituality of man's relation to nature," Stankard said, "and I develop illustrations to express that relation. I don't want my work to look labored. I want it to have an organic quality. If asked if it's craft or art, I have to think that ultimately the work itself will transcend any category."

Flameworking can be almost hypnotic, as the flame and the melting and the rod of glass going soft drains attention from the artist. So it is with Stankard. On the average, for every six pieces he makes, two will have to be destroyed because of distortion or bubbles in the glass. "In America," he said, "we have the luxury of being able to waste glass."

Although he lives in the most densely populated state in the nation, Paul Stankard gains ideas for his nature themes by walking in the woods near his home. He writes poetry, and of the poets he reads, Walt Whitman, he finds, has much to offer, for it was he who beseeched the saints of timeless verse: "Give me the splendid silent sun with its beams full-dazzling . . . Give me odorous at sunrise a garden of beautiful flowers where I can walk undisturbed."

Stankard's work aside, paperweights have moved through the history of glass with a measure of irrelevance. Collectors prize them, and there have been some pieces—crystal ones, especially—worthy of attention. For the most part, however, paperweights play in the kitsch league with lawn ornaments, crochet work that invokes blessings on the house, and souvenir pencils with miniature plastic sunflowers where the erasers are supposed to be. More than anything, I suppose, the glass paperweight is seen as the snow dome, the ubiquitous orange- or grapefruit-sized globe meant to be shaken in order to set the winter scene, with snow falling on, for example, the Capitol building in Washington, Baltimore's Harbor Place, or a Bavarian castle with two stags rutting by the moat.

Early glass paperweights were mastered by French glassmakers, one of whom was Baccarat, and they were both functional and decorative; they included the ones shaped like an egg and designed to be held as hand coolers. By the 1850s paperweights were being made in large numbers in the United States. Then, as with Paul Stankard now, they used flameworking to produce the pieces; more often than not, the decorations were of flowers or birds. In England many pieces were made in the shape of animals, including a famous lion one. Some may have held papers in place, but most were and continue to be assigned decorative duty, quite often on a shelf in the company of a porcelain figurine and/or a music box with mother-of-pearl inlay.

There are no flowers of glass more treasured in the United States than those at Harvard University. There are 847 pieces in the Ware Collection of Blaschka Glass Models of Plants, and they are each and together a ringing testament to glass as a material of inspiring beauty. Drawing close to a hundred thousand visitors a year, the collection is Harvard's top public attraction. They come mostly in the fall, when the foliage in Massachusetts and all of New England is itself like art on exhibit.

Two Germans created the collection that has been described as "an artistic marvel in the field of science and a scientific marvel in the field of art." In the late 1800s, in the town of Hosterwitz, not far from Dresden, Leopold Blaschka and his son Rudolph were casting replicas of marine life in glass. The director of the botanical museum at Harvard, George Goodale, came across the Blaschkas' work and decided that glass models of plants would be perfect for display in his museum, and a deal was struck with the father-son team. The Wares—Elizabeth and Mary Lee, her daughter— were the benefactors, a gift in memory of Elizabeth's dead husband, Charles, Harvard class of 1834. The work extended over many years, until Rudolph died in 1939. By that time, Harvard had in its collection the 847 models, each life-size and botanically correct. Included are 780 species of plants in 164 families. There is no single-subject collection of glass so thorough, so detailed, so stunningly lifelike, in all the world.

At first, the Germans used clear glass, which they painted. Later, after his father's death, Rudolph made his glass rods and tubes with the color in. They worked as Stankard works, the exacting, mesmerizing deliverance of artful form from flame-softened glass. In a description of Rudolph at work, Mary Lee Ware wrote, "You could only say that he waved a piece of glass in the air and the petal was outlined. Then a little pressure and a slight twist with pinchers and you had it."

Well, there's more to it than that. The glass artist working at the furnace directs his movements with measured control, not unlike the driving form of professional golfers or the subtlest movement of reins by riders of Vienna's famed Lippizaner stallions. They all make it look easy. But with glass, there are always hits and misses.

Chapter Fifteen

Glass and the Ballet of Light

Of all the genre of glass as art or craft, certainly none is more widespread and enduring as stained glass. It is found not only in great cathedrals, but also on the doors of shower stalls and on the shades for night-lights. Stained glass has also become an important architectural material, incorporating the glorious union of sunlight and colored glass in design.

But only in its ecclesiastical form does stained glass achieve its full grace and glory. It is then that the ages-old material becomes a benediction. Of all the stained glass in the world today, a full 90 percent holds religious themes.

The Abbé Suger, who in the twelfth century rebuilt the church of St. Denis outside Paris, was among the first to recognize the ability of glass to brighten mood and perception, to allow us to move outside our physical world, "urging us onward from the material to the immaterial." He knew that glass, colored and backlighted and positioned properly, appears as a beacon for the spirituality in the vast rises of a great basilica.

Much stained glass is made today as it was centuries ago.

Color is brought to the glass by adding metallic oxides to the basic mixture of sand, soda, and lime; the addition of silver, for example, will lend an amber color, as will real gold a pink-gold color. Pieces of the colored glasses are cut to fit a pattern—say, a life-size depiction of the Venerable Bede with book in one hand, quill pen in the other, as may be found at the American School in Regent's Park, London—and then joined with the use of lead cames, or rods, soldered together. The process demands not only the vision of art but also technical expertise. Good stained glass made today in this traditional way is expensive, as much as a thousand dollars a square foot.

After blossoming in medieval times, stained glass declined in popularity, then returned to prominence in the last half of the nineteenth century. William Morris, along with Edward Burne-Jones and Dante Gabriel Rossetti, all English artists and poets, brought forth the Arts and Crafts movement and a great surge of renewed interest in stained glass as art. The writer John Ruskin contributed heavily to the dialogue. For the first half of the century, the type of stained-glass windows being made was mostly fired enamel painting on a single piece of glass, requiring no lead pieces for joining. Enameled glass, Ruskin and others argued, compromised the transparency of glass and therefore its glorious play with light. Their voices were heard.

Glass as a medium for art and decoration was exalted in the second half of the century, for art nouveau, drawing its inspiration from William Morris and his designer friends, had come into vogue, and nothing so suited that style as glass. In America and across Europe, the great and ancient craft of stained-glass making regained its preeminence.

Even in Glasgow, the passions and strengths of good stained glass shone through the grime and soot of that cheerless Victorian city. In addition to the popularity of art nouveau, that surge of stained-glass-making and installation was caused by a spiraling rise in church construction following a wide schism within the Church of Scotland (through much of its history, the well-being of stained-glass-making has risen and fallen on the redemptive swells and ebbs of a populace getting right with the Lord after wars and other sinful diversions). Known as the Glasgow School, the elegant works of that time, from around 1870 to the start of the First World War, established the city as a leading center for the crafting of design in stained glass. From the depiction of small roses in the windows of Miss Cranston's Argyle Street tearooms to large panels given over to energetic harvest scenes, the production in Glasgow was far-ranging, but always tightly focused on the genius of it all.

In France, meanwhile, one of the greatest colored-glass artists of all times, Émile Gallé, was creating works of star quality in the art nouveau movement. His specialty was cameo pieces with two and three layers of glass, each of a different color, and each carved to achieve an overall image. Gallé's designs allowed his favorite depictions of animals and plants to be cast in reflecting and transmitting light, works of three-dimensional brilliance.

Great artists such as Matisse, Chagall, and Braque began working with the material, but it was an American, born in 1848, whose name for many came to stand for masterworks in stained glass. He was Louis Comfort Tiffany. His windows, lamps, and other pieces glow with translucent colors, and they stand among the greatest glassworks ever produced. He came along at a time when stained glass, as an expression of art, had been pushed into obscurity in the United States because of widespread disinterest.

Unlike most stained-glass artists, Tiffany was also in the business of making glass. The commercial colored glasses available to artists were not acceptable to him because their colors lacked vividness. He chose to make his own glass, and in 1885, a time of postwar revival of church building in the country, the newly founded Tiffany Glass Company began to share in the effort to meet the hungry demand for stained glass.

Louis Comfort Tiffany was one of the Tiffany jewelry family. He was wealthy and well-traveled, and imagination cadenced through his brain in quick step. He gave to art nouveau and to the world the works of a man driven by a fiery passion for innovative design. His specialty glass was opalescent—Favrile Glass—and with that he and his teams of workers (Tiffany did not blow glass himself) produced the lamps and vases and windows, all with the iridescence, design, and color that collectors have treasured to this day. In the stained glass of his windows, the waters and mountains and skies of landscapes seem not so much the handiwork of an artist as they do a superimposition from the outside. Tiffany glass was all mouth-blown, and, as a result, frequently flawed with the "seeds" and other imperfections of such glass. But he would turn those to his advantage, and in the end they would become like seals of authenticity. His glass had a weathered look, and that's the way he wanted it.

The lamps have come to be regarded by the general public as Tiffany's signature pieces, and they are indeed works to live through the ages. The lampshades were made of leaded glass, and, generally, they sat on bases of bronze. They varied widely in design: There was a prize-winning table lamp with eighteen lily-shaped shades of the sheeny Favrile Glass, and another of fiery reds and golds depicting maple leaves. The listings for lamps in the Tiffany Glass Company catalog was lengthy, and the prices high. Louis Comfort Tiffany is remembered now too because of a vase, a 1912 vessel called "Jack-in-the-Pulpit." With a slender stem and a mouth that flares out into a disk that droops on the lower front side, the ingenious design of the vase speaks to the art nouveau style as clearly as anything produced in glass during those thirty or forty years that straddled the turn of the century.

Technique was not Tiffany's strong suit, but that does not diminish the importance of it for what he did. He hired others to do the fabricating. His chief rival and the other great stained-glass artist of the time, John LaFarge, was more of his own man in the shop. For his contributions to the art of making stained glass, LaFarge deserves to be remembered more, but, alas, he came along at a time when Tiffany was destined to gather all the laurels.

And there were others working with glass during that time who would be remembered, though not for stained glass work: René Lalique, the French master of colorless sculptured glass, and Frederick Carder, who was born in England but spent most of his life working in the United States as a glassmaker and designer; unrivaled in his time as a craftsman working with decorative glass, he went on to cofound Steuben Glass Works. Both were giants in the glass arts scene of the first half of the twentieth century. During those years between the two great wars, they gave to the world works in glass so masterly in the use of color, design, and composition that they would become treasured.

The general conception of stained glass is one of a material in service to God. It has been that way for a long, long time, and it is doubtful if it will ever change. And why should it? Only as a glory of the church does stained glass attain its full potential. And that can be the glass of the famous windows in Chartres Cathedral, or that of a mission chapel in South Philadelphia. Before he died of the plague, at Tunis, Louis IX of France (1214–1270) did something for which he is remembered more than for having led the Seventh Crusade; he built the shrine of stained glass at

Sainte-Chapelle in Paris. The windows rise to fifty feet in height and surround the chapel on three sides. It is a monumental work not only in size, but in magnificence as well. More than just colored panels leaded together, Sainte-Chapelle's fifteen windows are of narrative glass, with more than eleven hundred scenes.

The thirteenth and fourteenth centuries, with Romanesque architecture out and Gothic in, were banner times for cathedral building. In addition to Sainte-Chapelle, there was Notre Dame in Paris, Cologne in Germany, Avila in Toledo, and Canterbury, Winchester, and Ely, all in England. And there was Chartres, a spired hosanna in glass and stone. Once a structure of Romanesque design, Chartres was rebuilt after a destructive fire in 1194 (of the cathedral's two spires, one is Gothic, the other Romanesque).

All together, there are 176 windows of stained glass in the vast embrace of Chartres's walls. There are windows of secular as well as religious themes, including forty or so that depict workers in various trades, such as wine making and meat cutting, that carried the economy of the medieval city of Chartres. There are other windows with pictures through which a narrative is strung, but it is in the galleries of saints and saviors and blessed events that the windows of Chartres dispense second thought to nonbelievers. Among those is the famed North Rose Window, a Gothic jewel in brilliant reds and blues. In the south ambulatory, Notre Dame de la Belle Verrière stares down in commanding but benevolent presence, her visage all the better for centuries of weathering.

The best stained-glass makers in all the world were in France at that time. The process they used seldom changed: with the design sketched on paper, the colored glass would be cut to patterns, and there would be some painting on various parts of the pattern of such details as an eye; after firing in a kiln to protect the enamel paint, the glass parts of the window were joined by soldering the lead cames together. They were highly skilled craftsmen working in what has always been a crossover field of decoration and architecture. Whatever they did, they did it to last, but there were wars to come, and nothing so challenges the permanence of a cathedral's stained glass as the turmoil of armed conflict.

During World War II, the windows of Chartres were removed—all of them in just seven days—and put away for safekeeping. That was not the case with the stained glass windows of another thirteenth-century church, this one in the French town of Rémy, about sixty-five miles north of Paris.

In the afternoon of August 2, 1944, eight American P-51 Mustang fighter planes out of a base in England appeared over German-occupied Rémy. Four dropped down to about two hundred feet to strafe an eighteen-car freight train standing at the small station under camouflage, while the other pilots continued high-altitude surveillance. Under fire, the train exploded; clearly, the boxcars were loaded with materials of powerful force, for the four Mustangs at ten thousand feet were wrenched from their positions by the shock wave of the explosion. Four hundred Germans were killed, as was an American pilot of one of the low-altitude fighters. A later investigation of the raid at Rémy concluded that the boxcars may have been loaded with warheads for rockets, or perhaps nitroglycerine, as well as ammunition.

And some of the stained glass windows in St. Denis, the church at Rémy, were destroyed—a small thing in the face of so many deaths, but something that the people there would not forget. "The people collected as much of the broken glass as they could find," said Stephan Lea Vell, a retired commercial airline pilot who lives in California. "There were seven windows damaged or destroyed."

Lea Vell has a hobby of researching military events, and, having read about the raid at Rémy, he decided to visit the town while on a trip to Paris in 1994. He wanted to visit the cemetery in Rémy where Houston Lee Braly, the American pilot who died in the raid, was buried. When his plane crashed, his body was recovered by the townspeople and kept hidden from the Germans after wrapping it in a parachute. In lieu of a headstone, they placed a blade from the propeller of his plane over the grave (the body has since been returned to Texas for reburial). "I was in the cemetery," Lea Vell recalled, "when an old lady on a bicycle came up and motioned for me to go with her. She took me to another old lady, who told her grandson to take me to the church. That's when I learned that the stained-glass windows were missing."

He saw an American flag hanging in the church, put there some fifty years earlier in celebration of the arrival of American troops. When the people learned that liberating Allied troops would soon be marching in, a woman among them, not knowing who would be the first to arrive, quickly sewed three flags—Free French, Canadian and American—using bedspreads for material. "The Betsy Ross story is a myth, but this one is true," Lea Vell said. "She put forty-eight stars in the American flag."

The church was central to the lives of the people there, as it had been to those who lived there before them, going back seven hundred years. In a French village like Rémy, seldom if ever caught up in a swirl of prosperity or sophistication, the commission by the church of the common rites of life, such as baptism and marriage, remains important. Stained glass is important, too, for that further sanctifies the church-based passages. For the people of Rémy, the destruction of the windows was a heavy loss.

In what one of them called a "debt of honor," the surviving pilots of the raid, along with Lea Vell, began a campaign to raise funds to have the glass replaced; both new glass and the old that was salvaged after the raid will be used in the replacement work. More than $100,000 was collected in the campaign, and as of mid-1997, all that remained to be done before the restoration could get under way was for the French government to give its approval. "The church is a historical monument, and you just can't do anything to a national monument without the government's approval," Lea Vell said. "We've heard that it may take two years to get that, but that's too long. We're going to push them on this."

When nations fight each other, stained-glass is left vulnerable to every explosive nuance of war. No one knew that better than the Reverend Harold Appleyard, who served as a major and chaplain with the Canadian Army during World War II. While in England, he collected fragments of stained glass that lay on the ground amid the ruins of churches and great cathedrals. He made a record of the origin of each piece. In almost all cases, he sought and received permission to remove the glass, but when that wasn't possible, it can be said that he invoked a prayerful right of eminent domain. He would not allow his vision for use of the glass to become clouded.

Major Appleyard had gone off to war from Meaford, a small town of four thousand in Ontario that sits on the shore of Lake Huron. There, he was the rector of Christ Church Anglican, a structure of modest size with a steeply pitched roof meant to stave off the snows of the harsh winters in that north country. It was there, he decided, that the collection of glass fragments would be preserved as windows as a memorial to those who died in the war, including the forty members of his regiment who lost their lives in France during and just after D Day.

He took his hoard to the English glazing firm of Cox and

Bernard, in Brighton, Sussex, and directed that windows be made of it. Recalling the incident for a newspaper reporter, a former employee of the firm said Major Appleyard insisted that the cement used in the procedure as a glazing putty to bond the glass to the joining lead rods be dried with sawdust rather than Talc-like powder, knowing that powder can damage old glass. The fragments were laid down in patterns with little continuity, meaning, for example, if there was only a partial depiction of a saint's head, the blank space would be filled in with whatever they had that resembled that head.

"You can tell it's broken glass," said Mrs. Merle Cooper of Meaford. She is treasurer of Christ Church, now with a congregation of 275 families. "But every piece is documented. We know where each came from."

There were enough fragments for four windows, including one containing stained glass only from the windows of churches designed by Christopher Wren. It is, by any standard of judgment, an extraordinary collection of stained glass, fragmented or not. It would be a hard search to find stained glass as old and as illustriously pedigreed as this anywhere else in North America.

For all of its rapturous affair with light, stained glass can be bad craft, and even worse art. For every "Virgin and Child" at Chartres, or Marc Chagall's "The Creation of Adam and Eve" at Metz, there are hundreds, thousands, maybe tens of thousands of offensive panels. And that's excluding the stained glass of hobbyists who burden relatives and friends with cargoes of blue whales and red hearts to hang on sliding glass doors.

But bad stained glass need not be small. It can be colossal, an overwhelming rejection of artistic honor. The stained glass in the world's largest Christian church is all of that. The Basilica of Our Lady of Peace in Yamoussoukro, a city in Côte d'Ivoire, in West Africa, rises in a scrub jungle of that sub-Saharan land like a mirage, and to come on it at night, when it stands as if caramelized in the light of its nearly two thousand flood lamps, sets the mind to chasing thoughts of folly most grand. And with the coming of sunlight, the thoughts are confirmed.

Imagine mouth-blown stained-glass windows that measure ten stories in height; imagine fifteen thousand panels—over twenty-five thousand square feet of glass—touched with four thousand different shades of color. And imagine all of that as just a part of an edifice incorporating seven hundred thousand tons of travertine marble, a basilica 525 feet tall (St. Peter's in Rome is 452 feet high) with enough pews of fine hardwoods to seat seven thousand. These are pews in which cool air rises just high enough from below to wrap a worshiper in a body bag of relief from the torrid equatorial climate. When crowds overflow onto the seven-and-a-half-acre marble-floored esplanade, the church can accommodate as many as 350,000 persons at a service.

Our Lady of Peace was offered to the Vatican, and the Vatican had to think a long time before accepting. Even now there is a strong hesitancy by Rome to acknowledge its weak sanction of the basilica. Perhaps the Holy See is embarrassed by the stabbing reality that a church believed to have cost anywhere from $140 million to $300 million flaunts its opulence in a country where the per capita income is less than the operating expense of the new digital organ for just one mass—in a country, too, where only 17 percent of the population is Catholic. It is no less embarrassing to have the man responsible for building the church depicted in one of the panels of stained glass, shown there in the company of Jesus Christ.

Félix Houphouët-Boigny was president of Côte d'Ivoire from 1960 until his death in 1993. He not only led the country to independence from France, but lifted its 15 mil-

lion or so residents to a level of relative—for West Africa—
economic comfort. But world coffee and cocoa prices col-
lapsed, and with them a large part of the country's economy.
By that time, Houphouët-Boigny was a person of extraordinary personal wealth, acquired in ways more of tradition
than shame. He could afford to buy immortality, and so he
always maintained that it was his own money that paid for
Our Lady of Peace. He would dismiss probing questions
about that by saying it was not important, that "God doesn't
need accounting books." His wish was that Africans from
across the continent would make their way to the church
on pilgrimage.

It is the head of Houphouët-Boigny, a convert to Catholi-
cism, shown in the stained glass at the feet of a Christ with
blond hair. That rather sums up the artistic level of the glass.
One critic saw it as a biblical narrative in comic-book style.
It is very colorful and massive, with halos like the wheels of
Conestoga wagons. Thought has to be given, however, to
what Our Lady of Peace would be like *without* the glass. If
nothing else, it is a colorful adornment on the chilling con-
structs of lunatic pride.

Stained glass, for the most part, requires restoration after
a time lapse of between 100 and 150 years. Among other
things, the enamel painted on the glass and then fired in an
oven starts to deteriorate. It is then that a person such as
David Fraser steps in. He is a master stained glass conservator, a person of rare talent and infinite patience.

You will not find David Fraser at work in Europe, minister-
ing to the windows of a twelfth-century cathedral, although he
could do that as well as any conservator. Rather, he is bringing
new life to the stained glass of windows in a church in Brooklyn. More than that, though, the windows in the St. Ann and
the Holy Trinity (Episcopal) Church on the borough's Mon-

tague Street contain an ensemble of nineteenth-century stained glass many experts consider to be not only of major importance, but a national treasure. They were created over a three-year period by the brothers William and John Bolton in commemoration of the church's opening in 1844.

The mid-1880s were a fecund time for the production of stained glass. "They were cranking out the windows then," Fraser said. "And of all of them, the ones here have the most historical importance. They are regarded as America's equivalent to the stained glass windows of Canterbury Cathedral." There are some sixty windows in all, amounting to seven thousand square feet of glass, and they are grouped on three levels, one of which depicts the lineage of Christ with a tree of Jesse, one the life of Christ from baptism to crucifixion, and the other scenes from the Old Testament. It is a sweeping and powerful glorification of Christianity.

As in-house master glass conservator, Fraser works with one apprentice and three interns. He himself trained under a master. The work at St. Ann consists of taking each window apart, piece by piece, and usually underwater as a precaution against contamination by dust. Sometimes the glazing that protects the glass has rusted, and that must be remedied. Cracks must be repaired, the lead cames strengthened. These windows have been jarred by the rumblings of subway cars under the church, and they have been damaged by air pollution.

"We do this work the same way it has been done for centuries," David Fraser said. "The only difference is that we have electricity for our soldering iron." He is aware that he must be on guard against causing adverse changes to the color, as has been done in other restoration projects. At Chartres, for example, a window lost depth of color during a restoration cleaning. There is usually close scrutiny of a restorer at work, since traditionalists seem to prefer that things be left as they are. The restoration at St. Ann has been in progress since 1979. Fraser predicts that it will take until the year 2002 to complete all of the windows.

If a person walked the streets of Brooklyn and counted all the churches, the total might come as a surprise. The borough that gave the world the personification of character of place happens to be crowded with churches—two thousand of them. But many have lost substantial numbers of members. St. Ann is among those. One of the Bolton brothers who crafted the church's spectacular windows is reported to have written, "I was wrecked once on the beautiful coast of color." No matter how empty the pews are on Sunday morning, that congregant of glorious color will always be in attendance in the church on Montague Street. David Fraser is seeing to that.

Of course, stained glass does not have to be inserted in the wall of a church. As an art piece, it can be put in a light box and hung on the wall like a painting. If that is done with stained glass made by certain of the artists in the field today, the effect can be riveting.

Judith Schaechter is such a stained-glass artist. She does not chronicle the life of Christ in her works, although she can become as devoted to a subject as any twelfth-century craftsman chained to the drama of Christianity. The novelist Rick Moody has written this about her: "Judith Schaechter has articulated a secular vocabulary for a medium whose first and most singular use was the expression of a very particular, unitary faith."

At first she was a painter, and she painted mostly cats, lots and lots of cats. But, moving to stained glass, she abandoned that for what at first were human figures twisting in some Stygian depth of violence and torture. Rape, arson, and murder were among the themes. Later, her work would become calmer, but no less on fire with emotion. "My work," she once said, "is not intended to make comfortable people unhappy. It's intended to make unhappy people comfortable." Obviously, Judith Schaechter, who is in her thirties, makes use of autobiographical expression in her work, for the imagery of the glass is much too haunting not to have been drawn, at least in good part, from personal experiences.

"I am not a religious person," Schaechter said, "but what I think is most intriguing about glass as an art medium—maybe science, too—is that it is not metaphorical in nature, but 'enlightenment embodied,' as the Abbé Suger, the medieval architect, said. Glass reminds me of rock-and-roll because it has an immediacy and a huge emotional impact. It's really powerful stuff. And it's beautiful. You can put a piece of crap in stained glass and it's going to look okay. Even when a piece of glass art *is* ugly, the viewer tends to be disappointed not so much by the ugliness itself as by the lost promise of beauty."

Judith Schaechter's work could not be as effective in any other medium. Stained glass carries the macabre with authority. The eerie beauty of the scenes in the glass she crafts—the colors, the lines, the form—preens itself in the gathered light. And there is a sense of martyrdom in many of her pieces, the imagery of heavenly atonement that plays so very well in stained glass.

She galls the hell out of some glass artists with didactic instincts when, for example, she is quoted in an art journal with this: "It is very hard to suggest a corpse in a work of art. Even in a photograph, a corpse just looks like a sleeping person. To make it dead, you have to make it a little grisly." For all of that, Schaechter has done much to get modern stained glass into the art galleries where the traditional view of the technique is one of craft rather than art. Art or craft? "Glass artists traditionally have had an inferiority complex because, again traditionally, it has been cooler to be an artist than a craftsman," Schaechter said. "I like to think my work is beautiful, and so I don't care if they consider it to be an art or craft."

Her techniques for making the windows are traditional, for the most part. The glass, of course, is mouth-blown into a cylinder, which is split, leaving the two sides to be heated and ironed flat. Rather than having the glass colored by adding metallic elements at the time of the initial melt, the glass is blown clear or with just a pale tint, after which the rich color is added in the form of a layered thin veneer. That is called "flash glass." A cartoon is drawn to serve as a pattern for cutting the glass, which she does using a steel wheel and a diamond grinder. Details are added by painting, to a large extent, but many of the effects can be achieved simply by engraving or sandblasting away the flash layer in certain places to allow the pale tint or clear glass to show through. If there is a need for shades of yellow, a form of silver sulfide is applied to the surfaces before the pieces are fired in a kiln at 1250°F. After that, the pieces are assembled and joined with lead cames. The binding together can also be done using copper foil, a technique first used by Tiffany. Backed with an adhesive, the foil is affixed between the pieces, and then the seams are sealed by soldering.

Across the country from the studio in Philadelphia where Judith Schaechter works, another woman is saying what it's like to make a piece of stained glass: "You work on it for so long without really seeing much, and then, all of a sudden, the light is coming through, and it's unbelievable. It's like giving birth."

Catherine (Cappy) Thompson's star as a stained glass artist has risen high. As an underboss in the Chihuly family of Northwest glasseteers, and an artist of a big talent that's forever evolving, she is paid wide attention by critics and gallery people. "The range of possibilities in working with glass is enormous, but I'm not interested in pushing the material," she said. "I just want to create a beautiful image."

Thompson's work is closely watched, for she has shoved aside all aspects of stained-glass making save for the painting of the glass with enamel, and that is now the focus of her work. Actually, it is a technique used as far back as the Middle Ages, known then and now as *grisaille*. She paints mostly on round vessels, first firing a drawing in black and gray on the glass, and then applying transparent colors, each of which is fired separately in the kiln. All of that is done by reverse painting on the inside of the vessel; in that way, being in the embrace of the glass, the narrative takes on added impact. She also paints with enamel on the outside of vessels of opaque, layered glass, using colors of intense richness.

The images are of primary importance in her work. They draw heavily on costumed mythology and fables. At other times, as Thompson once wrote in an introductory program for an exhibit of her work at the Seattle Art Museum, "I work more intuitively, building my own narrative in a kind of picture-poem based on my imagination and the use of symbols."

In explaining what she does, and why, Cappy Thompson, like Judith Schaechter, does not often wander into that thicket of incomprehensible babble that comes as second nature to many makers of glass art and those who review it. She has learned her craft as well as any person who works with stained or painted glass today. She's not a fast worker— fifteen pieces or so a year—but when there's something cooking in the kiln in her Seattle studio, it's something very good.

Down through the centuries, stained glass has been essentially in the service of architecture. Whether for a cathedral window or a panel in the glass canopy of a mall, the material is meant not only to gather light unto itself, but to have the effect set a mood for the structure's interior. But even with that, architects have not looked kindly on stained glass through most of the twentieth century. They have dismissed it as being either too kitschy or too ecclesiastical. But soft stirrings of change have been detected, and it is not unthinkable that a strong swing to renewed use of stained glass in architecture may be underway by the end of the new century's first decade.

"In the 1980s, requests for stained glass in the house were rare," Dean Brenneman, an architect who practices in Rockville, Maryland, said. "But that has changed now, and many more clients are interested in the material. They want it for door panels and windows, especially in privacy areas. They are interested now in engraved glass as well."

Whatever happens now, stained glass, like the cocker spaniel, no doubt will continue its pattern of rising and falling in popularity. The time will come again when panels featuring purple wisteria and snowy egrets are in use in the home. Seldom does stained glass help boost the appeal of a moderately priced house, in that it becomes pretentious and serves only to emphasize the drabness of the place. It is best to limit use of the material in such cases to something in the bathroom, preferably a window of hexagonal shape.

Frank Lloyd Wright championed the use of stained glass in architecture, but not glass of busy design and colors too thick to take the light properly. The famed architect was using stained glass in his Prairie School designs almost from the time he began his private practice in 1893. The designs of color in his windows are geometric—clean, linear patterns set on clear glass. It was exquisite stained-glass design, ahead of its time, and much of it remains today in the houses and the works of other commissions he completed before his death in 1959.

Wright, of course, customized his windows and door panels for the individual client, and, since most of the glass is clear, for the setting of the structure. But such exclusiveness, caught in the sweep of mass marketing, has vanished: the Anderson Window Company of Bayport, Minnesota, offers reproductions of the Wright windows in four different designs. Also on the market today is an acrylic overlay meant to be taken for stained glass. To install this, lead strips are cemented to the glass in the pattern of the desired design. Pieces of the colored acrylic are then cut, fitted between the strips, and glued to the glass.

Instant hibiscus.

Chapter Sixteen

The Music of Glass

The man is like a halcyon, the way he has a calming effect on the second- and third-graders seated on the floor before him. He stands at a table on which there are dozens of glasses, all brandy snifters, and if you ask him why they are there—why he has laid out at least fifty glasses of many sizes, and anchored them all to the table with rubber bands— he may reply that he did it so the children could hear the angels sing.

When Jamey Turner works with glass with his hands, the voices rise in exultation, and fade in solemn lament, but always grace on the ear with heavenly resonance. For him, glass should be not so much restructured by science as caressed by those with music in their hearts.

I met Turner, a pixyish man with boundless enthusiasm for almost everything, as he was preparing to make music for the students, to run his fingers over the rims of the glasses until such offerings as "Ode to Joy" from Beethoven's Ninth Symphony rang through the room.

The instrument Jamey Turner plays is called the glass harp, and he may well be the most accomplished glass-harp player in the world. Of course, it is almost a ritual of life to sit at the kitchen table at one time or another and produce a sound by rubbing a finger over the rim of a glass. It wasn't a long-lasting interest of most people's early years, but rather something on a par with scratching fingernails on a blackboard.

Turner, however, heard something in that voice of glass that most of us missed at first. Johann Wolfgang von Goethe, the German poet and dramatist, heard something special in it, too—something he called the "sound of the universe." By the time Herr Goethe got around to starting to write "Faust," in the early 1880s, the glass harp was already at least a hundred years old. In 1746, another notable German, the composer Christoph Gluck, appeared in concert in London, and devoted the entire program to playing the glass harp. Benjamin Franklin first came across the instrument in England, and, as invention was the statesman's wont, he devised a mechanical version of the harp in which a treadle is pumped to turn the vessels against a rubbing device.

Few materials respond so musically to the human touch as glass. Once, on a small and very remote island in the Pacific, I watched and listened as an aged sexton struck a large, empty steel canister with a club made of the wood of a monkeypod tree, to summon the islanders to church services (the Methodists had gotten to them many years before), and that was a good sound. And something should be said for the glockenspiel and xylophone, but again, as with the steel canister, each of those must be struck by an object such as a hammer. With the glass harp, it takes only a human touch—fingers on the glass, the right touch on the right place, and, as is common with such contacts, wails of pleasure in response.

It is not that easy, of course, just as it is not easy to play the musical saw, which Jamey Turner also does brilliantly.

In preparing for a performance, the glass-harp player, unlike, say, a cellist, must spend an hour or so just setting up the instrument. There must be a table to act as a sound board, but not just any table. "A boat builder made my table," Turner said. "We tested two dozen different types of wood before we got the sound we wanted." It is important to him that the sound be perfectly pitched, for his sensitivity to tone quality is acute.

He may then set up as many as sixty glasses of all sizes on the table. They are not crystal, but brandy snifters made of regular glass. Crystal produces an elegant sound, but it is a sound extremely difficult to control in that it takes too long to die out, and that will not do, of course, for passages of allegro tempo. Jamey Turner will tell you that one of the great glass-harp players of all time, the late Bruno Hoffman, used crystal and had trouble with sounds that tarried too long.

Turner is not a part-time musician who comes to a performance from a day job as a telephone lineman or produce manager at the Safeway. He is a full-time musician, with a glass-harp specialty, who travels the world filling concert engagements. He has performed on the glass harp with the National Symphony at the Kennedy Center in Washington, and with the Philadelphia Orchestra, the New Orleans Philharmonic, and other large ensembles. He has rubbed the rims at the White House, the Library of Congress, on all the late-night television shows, and at New York's Lincoln Center with Beverly Sills. Londoners have heard the ethereal sounds of his glasses as they filled the three-hundred-year-old Christopher Wren–designed Church of St. Anne and St. Agnes; he played Mozart and Vivaldi there, and those who came to listen were filled with fascination and joy.

Some teachers joined the young students in the Fairfax schoolroom, drawn there by the intrigue of being entertained by someone who has been billed as "the Horowitz of glass." But first, the glasses had to be tuned by putting water into them. Each snifter had to hold an exact amount of distilled water (the more water, the lower the pitch). Turner had not eaten ice cream for two days before this performance, as the dairy food tends to alter the texture of the skin of his fingers and thus upset his touch on the rims. "If I forget and do eat ice cream, I know how to counteract the reaction, but even so, my touch is still a little off," he said.

Jamey Turner trains for his sessions with the glasses. What he eats, he says, is a key element to his success before an audience. There are times when his fingers are a bit squeaky—a condition of no less hindrance than the tennis elbow of a cymbalist—and he will know right away that he must increase his intake of milled rice.

So, with enough rice and controlled intakes of mocha chip, his virtuosity with the instrument is assured as the pressure and tempo of his finger movements raise the voice of glass. The sound was particularly good on this day before the students, and later the assessment was made that the performance was more popular than recess. Before ending the program, Turner pushed the table to the side in order to have room to play a few selections on the saw, and finally the wrench harp (wrenches should not be thought of only in terms of nuts and bolts; the rich sounds of wrenches of different sizes being struck by a hammer are suitable for music, especially for something with the solemnity of a requiem).

There have been satellite recordings of sounds coming from the planet Jupiter, and, eerily, they are similar to those of the glass harp. I suggested to Jamey Turner that maybe there is another, far away, who shares his talent with the instrument. He said he hopes so.

It would be good to think that children at the schools where Jamey Turner performs will grow up with memories of his musical magic with glass cemented in their minds. Indeed, what are the lasting memories of preteen associations with glass? Soaping the windshields of cars on Halloween; a fishbowl overturned and guppies flapping around amidst the shards like moths around a porch light; a bat, a baseball, a broken picture window.

And this: the feel of a glass marble against the cocked thumb, about to be fired into a circle drawn in the dirt, powered by all the tongue-biting competitive drive a marble-shooter can muster up. One on one, and no marble moms on the sidelines. Soccer be damned—it's marbles that plant the seeds of character and leadership in the presidents and CEOs of tomorrow.

The largest maker of glass marbles in the country is in Paden City, West Virginia. There, where skinned knuckles among the young of the world are viewed as economic indicators, a company called Marble King manufactures one million glass beauties each day. Beri Fox, business manager, said a resurgence of marble playing in the United States occurred around the mid-1990s, and the reason is not clear (no concurrent rise in mumblety-peg popularity was reported).

It was in the 1920s and 1930s that toy marbles rolled across the land in record numbers. Playing the game was a childhood pleasure of the highest order. At the time there were as many as a hundred glass factories along the Ohio River from Wheeling, West Virginia, to Cincinnati.

As often happens with good times, however, they go bad, and by the 1950s the glass marble industry in the United States was suffering. And then the Japanese, in what must have been one of their first postwar recovery successes, introduced a glass marble in America that won favor with youngsters wherever marbles were played. It was the design—the cat's-eye—that the shooters liked. Indeed, it was about the only design they would buy, and so Marble King set out to duplicate it.

It took five years, but the West Virginia firm did succeed in marketing a cat's-eye; today the design remains as the flagship of the inventory there. Techniques for making the cat's-eye and other richly colored designs have improved over the years. When the toys were first made, blown by mouth, the "round" shooters not only had bubbles and other flaws, but corners as well. Today, at Marble King, production is automated, using a system whereby molten glass sags through holes in a moving belt to form the spheres.

Almost all of the glass used by Marble King for making marbles is cullet, or recycled glass. Quality control is high at the company's plant, and while the rejection rate is small, chipped marbles and others less than perfect are pulled before packaging and routed to the cullet pile. As Zlotnick the furrier once had a large stuffed polar bear on the sidewalk in front of his store on G Street, in Washington, D.C., so too does Marble King make a statement with design at the plant's entranceway. Rising there for six or seven feet is a mound of marbles, most of them chipped or otherwise flawed, but perfect for rolling through memories of old shooters.

Marble King manufactures for mass distribution. The company's marbles are not expensive, and the usual method of buying them is by the bagful. There is another company, Gibson Glass, in the West Virginia town of Milton, that caters to serious collectors. Gibson marbles can cost more than $10 for just one. "The marbles we make are not for the kid who goes to the store and buys them by the bag,"

Charles Gibson, the owner, said. "Those are machine-made. We make only three thousand to five thousand marbles a year, and they are made by hand. That's still a lot of marbles for the kind we craft."

Gibson uses a lime-based crystal for its marbles, and with the signature design being one of snakeskin you have to think that a city tower sheathed in glass of that design would set the tone for new urbanscapes in our time. The designs and colors of glass marbles have been like mosaics on the playgrounds and neighborhood backyards of this earth. It is an extension of the long history of colored glass, starting with the beads made by the ancient Egyptians, and going on to be strung as a rosary, or positioned as a navel adornment for a belly dancer. Meanwhile, the National Marble Tournament, started in 1923, continues to be held each year at Wildwood, New Jersey.

Jamey Turner concluded his concert at the school in Fairfax, Virginia, and said he thought it went well. His few selections on the musical saw were well received, as was a surprisingly melodious and tender offering on the wrenches. But it was the glass harp that won their hearts—that had 30 kids smiling and feeling good. And some of the teachers, having dislodged their minds from the vise of indifference to the materials that serve us so well in life, said they had no idea glass could be made to make such nice music. "Glass!" said one incredulously. "Imagine that."

Imagine this: It doesn't even cast a shadow.

PART IV

REFLECTIONS

There are two ways of spreading light; to be the candle or the mirror that reflects it.

—EDITH WHARTON

Chapter Seventeen

The Mirror

In my reverie:

The man, in his twenties, enters the elevator on the twelfth floor and quickly makes his way through a clot of passengers to the rear, where he turns to look directly into the mirror covering the top half of the wall. He becomes oblivious to all but his reflected image, and so he cares little that his vanity starts to carry through the car like poison gas, killing what sweetness of nature remains among the others aboard. Up close to the mirror, he gives a soft touch to the sides of his hair, draws back his lips for a look at his teeth, and swivels his head from side to side for profile inspection. By the time the elevator reaches the ground floor, the other passengers, out of control with rage, have pummeled him to the floor with briefcases and purses; each man and woman, upon departing, reaches down to muss his hair.

What if all the mirrors suddenly vanished from the earth, withered to nothingness under the ceaseless staring of the narcissi? Would we then return to the calm waters where man and woman probably first saw images of themselves, the flat forms of the faces lined with every ripple in the water, the expressions drawn tight where caught in surprise?

No doubt we would do most anything to find a replacement reflective surface, for the glass mirror serves our insatiable hunger for visual self-approval. Most people like to look at themselves, if not because they think they are attractive, then because they want to see if things have improved since the last look. Noble cheekbones or acute overbite, the mirror throws back what it gets, and that can be heartening or devastating for the one standing before it. As George Herbert, a seventeenth-century English writer with a wide frame of reference to glass in his works, once wrote, "What your glass tells you will not be told by counsel."

A magical moment:

It is morning and a woman sits before a mirror attending to coiffing matters when a figure appears behind her, a man about to leave the room. Without turning, she smiles at his image in the mirror, he smiles back, and in that way, with all that crisscrossing of reflections, they dedicate the moment to the passion of the night before.

The mirror has always been associated with magic. Philosophers, ancient astronomers, mystics, and shamans—all have turned to the mirror for aid in divining their self-proclaimed truths. They even had it that the moon is a mirror, reflecting earth and all its features.

It was not a passing fancy, but rather one that lasted for many years and drew the attention of the likes of Plutarch and Leonardo da Vinci. There were mirrors, some believed, that could predict the future, that could tell if a person's mind was cluttered with lustful thoughts, that could draw the soul from the body and have it appear in tenuous reflection.

And the curse of the mirror when cracked is a superstition older than the mirror itself, going back to the times when reflections were captured in devices made of polished metal. In one interpretation of the curse, it was a rather widely held belief among the ancient Romans that the average person went for seven years without a major change in health, and when a mirror was broken, it was taken as a sign that the next seven-year cycle would be full of bad livers, congested lungs, and general ill health. A more plausible explanation had to do with economics: mirrors were expensive, and the threat of bad luck was a way to have the servants and maids take added care when handling the pieces.

And it's still with us today, that blackest of black cats, the broken mirror. Seven years of bad luck, and who can deny having felt a pinch of apprehension at the time? Alfred, Lord Tennyson, had it right in "The Lady of Shalott":

> She left the web, she left the loom,
> She made three paces thro' the room,
> She saw the water-lily bloom,
> She saw the helmet and the plume,
> She looked down to Camelot.
> Out flew the web and floated wide;
> The mirror crack'd from side to side.
> "The curse is come upon me," cried
> The Lady of Shalott.

With its investiture of magic, the mirror has intrigued mankind from the beginning. Some cultures turn their mirrors to the wall at times of storms, for fear they will attract lightning, and family deaths, when they might snatch the soul away. It is not beyond their beliefs that a mirror could do those things, not when they knew Prometheus used the reflective accessory to steal fire from the gods.

Christianity held a suspicion of the mirror when it was in the hands of a woman, for in the eyes of believers, it became an instrument of evil—a scepter for the monarchs of vanity. Although that papal prerogative for finding the devil in the damnedest places is more restrained now, the mirror as a symbol of lust and self-importance can still be seen through the representations on the stained-glass windows of some major cathedrals, including Notre Dame.

We do not assign such traits of the supernatural to other accessories, but then the umbrella or pocket comb cannot confirm the suspected presence of a nose shining like a beacon. As wonderful as it is, a Rolex watch, unlike a mirror, cannot unmask the insecurities of those for whom beauty is a meal ticket: Five years after her death in 1992, it was revealed that Marlene Dietrich, to avoid seeing herself in daylight, when little can hide the ravages of advanced age, had only smoked mirrors in her apartment on Park Avenue in New York City.

It was nothing more than a mirror that a woman telemarketer, new on the job, turned to for sales training, as related in the *New York Times* of March 24, 1994:

"Her first week . . . was a disaster as she tried to persuade people to change long distance companies. People either hung up on her or took the time to tell her off. . . . It was at that point, she recalled, that she turned to a mirror for some role playing.

" 'I acted like I was the customer,' she said. 'I asked questions and answered myself back. I used my rebuttals, and I felt real confident.'

"So she took the mirror back to work.

" 'It was excellent because it was a very hard program, and I reached the goal in one day,' she said. 'I've been carrying my mirror ever since.' "

Before mirror makers turned to glass, they used stone, and metals such as bronze, brass and silver. All were highly polished, of course, but, still, the facial images cast on those surfaces must have deflated a lot of swell imagoes. The first glass used for a mirror was natural volcanic glass, the shiny black obsidian. It wasn't until the fourteenth century that man-made glass was used, and that was in Venice.

Once again Venice ruled over a sustained run of glassmaking, and the fine mirrors of crystallized glass made in Murano set the standard for the world. At that time, the mirror was not so much a wall adornment as it was a small, decorative personal item. It sat on the dressing table or was carried on the person. There were no great halls of mirrors yet, no mirrors for ballerinas at the barre. All of that would have to wait for the development of technology to produce larger sheets of glass—plate glass.

It was then that the mirrors of grand design could be made more easily, and none were so lavish as those crafted in France. For Louis XIV and other seventeenth-century French royalty, much of life passed under the reflective eye of the mirror. The Sun King filled Versailles with mirrors; in the Hall of Mirrors of the palace, there were mirrors on a line of seventeen arched doors, each curved at the top so that the richly framed glass flowed across the room in waves.

In other rooms of other palaces, mirrors were placed so that they captured the images of each other, thereby filling the spaces with glassy paths that never ended, never converged. There were mirrors set in the paneling, and mirrors took the light of candles and made a sun's dawning of it.

With the onset of the French Revolution in 1789, the mirrors of royalty, many set in frames of gold and silver and precious stones, soon belonged to the people. They were sold and otherwise moved out of the halls and great rooms.

So it has never stopped, this obsession with the glass treated to reflect. We go to the movies and watch actors staring into mirrors while contemplating murder, marriage, or dieting. If they're not doing that, they're telling us they can't have an affair, or refuse to donate a kidney to an ailing brother, because "I have to look at myself in the mirror each morning."

"Throughout the years," the late humorist Erma Bombeck once wrote, "a woman discovers there are kind mirrors and mean-spirited ones." Again, the mirror is glazed with intimations of mortality.

The basic technology for making the common mirrors in use today involves coating one side of a piece of plate glass with a reflective material. In the early eighteenth century, when mirror making was moving ahead to become an important industry in Europe and the United States, tinfoil with a covering of mercury was often used as the backing, and it wasn't until a century later that silvering was introduced.

The plate glass used now for mirrors is made by having the molten glass float on a bath of tin. In order to give an image without distortion, the glass must be free of bubbles and other flaws. With silvering, now refined and still in wide use, one of the sides of the glass is coated with silver nitrate mixed with other elements. The idea is to have the silver nitrate take a crystalline metallic form, and that is done by adding a reducing agent (dissolved glucose works well) to the coating, which, as a final step, is covered with lacquer. In a new technology announced in 1996 by the British glassmaker Pilkington, a silicon coating can be used for the backing.

There are specialty mirrors, of course, such as the ones used in optical applications, and crafting those can involve much more complex technologies. In some cases there, the melt must be done in clay pots, and the reflective coating applied through flash vaporization of certain metals; it is that vapor of, say, aluminum, that serves as the lustrous backing.

Had today's reflecting technology been known in biblical times, it may not have come to pass that Corinthians would have us in the position where "for now we see through a glass, darkly." In one revision, it is "Now we see but a poor reflection as in a mirror," and in another, "For we see in a mirror dimly." Either way, it's the gospel truth that, but for a good backing of shiny silver, we would be deprived of a snippet of scripture worn in our vocabulary like a favorite broach.

Not all mirrors give a true image, of course. The mirrors in the amusement park's funhouse throw back grotesque distortions, squeezing us down into puffy toads and stretching us tall and thin. This all has to do with the shape of the glass and how it affects the way light rays reflect from it. With the convex mirror, for example, the rays reflect away from each other, making the image appear smaller than it is. With the concave spherical mirror, parallel light rays are reflected toward each other, making the object seem larger.

The automobile rearview mirror with a feature to block the glare of the bright headlights of a tailing car contains dual reflective surfaces. Positioned with the push of a lever, the nighttime glass reflects only about 4 percent of the light, while the daytime glass takes almost all of it.

But the utility of a rearview mirror goes beyond that. At the stoplight, it becomes a looking glass for the application of makeup. In the taxicab, it serves the driver as a monitor for the deportment of passengers. And of course the rearview mirror announces the presence of the lights of a police car, which may or may not be welcome, but is there just the same, with the incessant flashing bouncing bursts of bright colors off the mirror.

The time has passed, I suppose, when anthropologists may yet come upon a lost tribe never with the benefit of a mirror or mountain pool in which to see themselves—a tribe in which only the eyes of others know the appearance of each member. It would be useful to deprive members of a study group the use of a mirror or any other reflective surface for an extended period of time—a year, say—and see how they fare. Would the narcissistic among them become weaned from their vanity? Would the unattractive, relieved of painful self-viewing, become convinced that there are now dimples where there were none before, that the bulbous nose, though unseen now, feels finely pinched, or grand as that of a Roman?

If nothing else, such a study would highlight the dependency of us all on pieces of reflecting glass. And, perhaps subconsciously, we see in the mirror not only ourselves, but ourselves as we relate to those around us. We look in the mirror in the morning after a bacchanalian night and see not only our bloodshot eyes, but also the ghost of the clerk who sells us Visine. The mirror carries us through life on a roller coaster of confirmation and denial, and, by its hold on us, straps us in for the ride.

Mirrors, of course, are not just for self-imaging. They are used in telescopes and microscopes, for solar systems and the sheathing of buildings. They are also made as decorative pieces for the home—panels of mirrors on the walls to make a room seem larger, mirrors over the fireplace, and elaborately framed mirrors hung among works of art.

Serge Roche, in his definitive book, *Mirrors,* quotes one Henri de Regnier, who was writing in 1931: "Everything seems to conspire to make our picture-hungry generation delight in the reflected Image. Perhaps the mirrors in our homes will be ever-open doors for dreams."

Dream away, but be prepared to know that the mirror does lie. It lies if it bulges or sags in the middle, and it lies if the coating on the back is badly scratched. Mostly, though, the glass mirror lies when it reads the printed word. That doesn't mean that if you say "tomato" the mirror will say "tomahto." No, if you give it *tomato* it gives you back *otamot,* which could be an Indian, as far as a lot of us know.

But we forgive the mirror for its spelling faults. It is much too involved with the bigger picture.

Select Bibliography

Arwas, Victor. *Glass/Art Nouveau to Art Deco.* New York: Rizzoli International Publications, Inc., 1978.

Beard, Geoffrey. *International Modern Glass.* New York: Charles Scribner's Sons, 1978.

Berlye, Milton. *Encyclopedia of Working With Glass.* New York: Dodd, 1983.

Blonson, Gary. *William Morris—Artifacts, Glass.* New York: Abbeville Press, 1996.

The Builders—Marvels of Engineering, Elizabeth L. Newhouse, editor. Washington: National Geographic Society, 1992.

Charleston, Robert J. *Masterpieces of Glass: A World History from the Corning Museum of Glass.* New York: Harry N. Abrams, Inc., 1990.

Chihuly Over Venice: Part 1: Nuutarjävi, Finland Portfolios.

Photography by Russell Johnson. Seattle: Portland Press, Inc., 1997.

Dale Chihuly Seaforms. Introduction by Sylvia Earle. Seattle: Portland Press, Inc., 1995.

Dale Chihuly Venetians. Introduction by Ron Glowen. Altadena, California: Twin Palms Publishers, 1989.

Dawes, Nicholas M. *Lalique Glass*. New York: Crown Publishers, Inc., 1986.

Duncan, Alastair. *Louis Comfort Tiffany (Library of American Art)*. New York: Harry N. Abrams, Inc., 1992.

Elliott, Cecil D. *Technics and Architecture*. Cambridge: The MIT Press, 1993.

Frank Lloyd Wright In His Renderings 1887-1959. Yukio Futagawa, editor, text by Bruce Brooks Pfeiffer. Tokyo: A.D.A. Edita, 1984.

Frantz, Susanne K. *Contemporary Glass*. New York: Harry N. Abrams, Inc., 1989.

Gardner, Paul V. *Frederick Carder: Portrait of a Glassmaker*. Corning, New York: Corning Museum of Glass, 1985.

Grover, Ray. *English Cameo Glass*. New York: Crown Publishers, Inc., 1980.

Hajdamach, Charles R. *British Glass, 1800-1914*. London: Antique Collectors' Club, 1992.

Huxtable, Ada Louise. *The Tall Building Artistically Reconsidered*. New York: Pantheon Books, 1984.

Klein, Dan. Glass: *A Contemporary Art*. New York: Rizzoli International Publications, Inc., 1989.

Klein, Dan, and Lloyd Ward. *The History of Glass*. London: Macdonald & Co. Ltd., 1989.

Kovel, Ralph and Terry. *Kovel's Depression Glass and American Dinnerware Price List*. New York: Crown Publishers, Inc., 1991.

Lilyquist, Christine, and R.H. Brill and M.T. Wypyski. *Studies in Early Egyptian Glass*. New York: Metropolitan Museum of Art, 1993.

Lindsey, Bessie M. *American Historical Glass*. Charles Tuttle, 1967.

McKearin, George. *American Glass*. New York: Crown Publishers, Inc., 1948.

Mehlman, Felice. *Phaidon Guide to Glass*. New York: Prentice-Hall, 1983.

Miller, Bonnie J. *Out of the Fire—Contemporary Glass Artists and Their Work*. San Francisco: Chronicle Books, 1991.

Oldknow, Tina. *Pilchuck: A Glass School*. Seattle: University of Washington Press, 1996.

Pfaender, Heinz G. *Schott Guide to Glass (Second Edition)*. London: Chapman & Hall, London, 1996.

Phillips, Phoebe. *The Encyclopedia of Glass*. Crescent Press, 1988.

Ruffner, Ginny. *Glass: A Material in the Service of Meaning*. Seattle: University of Washington Press, 1992.

Stanislav Libenský/Jaroslava Brychtová. A 40-Year Collaboration in Glass. Susanne K. Frantz, editor. Munich: Prestel-Verlag, With the Corning Museum of Glass, 1994.

Tait, Hugh. *Glass: Five Thousand Years*. New York: Harry N. Abrams, Inc., 1991.

Woods, Mary, and Arete Warren. *Glass Houses*. London: Aurum Press Ltd., 1988.

William Morris—Glass, Artifact and Art. Kate Elliot, editor. Seattle: Distributed by The University of Washington Press, 1989.

Zerwick, Chloe. *A Short History of Glass (Second Edition)*. New York: Harry N. Abrams, Inc., 1990.